Social Change in Southeast Asia

Social Change in Southeast Asia

edited by Johannes Dragsbaek Schmidt, Jacques Hersh and Niels Fold

 LONGMAN

Addison Wesley Longman Limited
Edinburgh Gate
Harlow
Essex CM20 2JE
United Kingdom
and Associated Companies throughout the world

*Published in the United States of America
by Addison Wesley Longman, New York*

First published 1997

ISBN 0–582–31734–7

British Library Cataloguing-in-Publication-Data
A catalogue record for this book is available from the British Library

Library of Congress Cataloging-in-Publication Data

Set by 3 in 10/11pt Palatino
Produced through Longman Malaysia, PA

Contents

Preface

This collection of contributions by international specialists on Southeast Asia is the result of collaboration made possible through the Nordic Association for Southeast Asian Studies and support from Research Centre on Development and International Relations at Aalborg University, Denmark.

The intention of the editors of this volume is to draw attention to the evolution of social relations in a region of the world which finds itself in a dynamic process of economic growth and transformation. The volume is divided into three comprehensive parts around the thematics of the internal and external dynamics behind social change in Southeast Asia; the issue of classes and development, culture and political legitimation, the industrialization process and the question of labour regulation. The editors have written a substantial introduction and conclusion which problematize these emerging 'miracle societies' and draw some international and general perspectives, theoretically as well as empirically. These two chapters emphasize the need for a reorientation in building theories and the way scholars approach the subjects involved.

The contributions have been commissioned and written so as to reflect as far as possible the variety of substantive subjects involved. Nine of the chapters were presented in an earlier form to a conference on 'Emerging Classes and Growing Inequalities in Southeast Asia' held in Denmark. All have been substantially rewritten, and some completely recast, in the light of the debate and critical comments which emerged from subsequent exchanges.

While the chapters differ markedly in their focus and approach as well as cases, they each constitute an attempt to rethink and reintroduce the nexus of social change taking place in Southeast Asia. The extent to which the state is able to manage the process of transformation through its influence on the behaviour of society can be grasped by looking at the economic, political and cultural spheres.

Economic growth and industrialization by definition transform the given societal foundations and give rise to new constellations of interests. In the final analysis these reflect the social composition of these dynamic societies. In Southeast Asia, the emergence of a middle class and the ideology of consumerism constitute a novel element in the strategy of late modernization. This creates tensions as to how to stage-manage the relationship between the different actors: the state, the market and civil society.

Each individual chapter approaches the problem from the point of departure of the author's interest. Thus, themes such as democratization, the role of the middle classes, social engineering, the political economy of industrialization, culture and consumerism, which all reflect the reality of present-day Southeast Asia, are implicitly or explicitly discussed by the contributors.

Acknowledgements

The editors would like to acknowledge the financial support from the Danish Social Science Research Council, the Council of Development Research and the Department of Development and Planning, at Aalborg University.

We express grateful thanks for critical comments from the Chair of the Nordic Association for Southeast Asian Studies, Ingela Palmgren, Department of Economic History at Lund University, and colleagues at the Research Centre on Development and International Relations, Li Xing and Susanne Thorbek. Addison Wesley Longman's Matthew Smith's encouragement is highly appreciated, and stimulating ideas from three anonymous reviewers have been taken into consideration in the final revision of the book.

Furthermore, important and valuable suggestions were given by the following: Kamaruding Absulsomad, Bundhit Pien-Ampai, Anne Jerneck, Leena Örnberg and Xiaolin Pei, Lund University; Rune Tjelland PRIO, Oslo; Thommy Svensson and Hans Antlöv, Nordic Institute of Asian Studies; Anne Karen Böcher, Anne Marie Skovsgaard, Lotte Thomsen and Jakob Lindahl, Institute of Geography, Copenhagen; Göran Lindgren, Department of Peace and Conflict Research, Uppsala; Monica Lindberg-Falk, Department of Social Anthropology, Gothenburg; Hedvig Brorsson, Department of History, Stockholm; Lars Kjärholm, School of Social Anthropology, Aarhus; Daniel Flemming International Development Studies, Roskilde; Marja-Leena Heikkilä-Horn, Institute for Historical Religion, Aabo; Erja Kettunen, Helsinki School of Economics.

The patience and continued support of Ellen Nyrup are highly appreciated. Without her professional skills and good sense this manuscript would never have been improved. We would also like to thank Morten Nejsig Olesen, Yao Christensen and Tonny Andersen for their help at various stages of the project.

As always, all ideas and criticisms from any of the above mentioned scholars and friends are totally absolved from blame for the final result in this book. It is, of course, the sole responsibility of the editors and individual authors.

We are grateful to the following for permission to reproduce copyright material:

The World Bank for tables 2.1, 2.2 and 2.3 from *Sustaining Rapid Development in East Asia and the Pacific.* Development in Practice Series. The World Bank 1993; Dr Dang Duc Dam for tables 8.1, 8.2, 8.3, 8.4 and 8.5 from Dang Duc Dam – *Vietnam's Economy 1986–1995*, The Gioi Publishers, Hanoi 1995; Professor Gerry Rodgers and The International Institute for Labour Studies for figures 7.1 and 7.2 from Boyer in Gerry Rodgers (ed) *Workers, Institutions and Economic Growth in Asia*, International Institute for Labour Studies, Geneva 1994

Whilst every effort has been made to trace owners of copyright material, in a few cases this has proved impossible and we would like to take this opportunity to apologise to any copyright holders whose rights we may have unwittingly infringed.

Abbreviations

AFP	Agence France Press
AFTA	ASEAN Free Trade Area
ARDP	Accelerated Rural Development Program (Thailand)
ASEAN	Association of South-East Asian Nations
ASEAN-4	Association of South-East Asian Nations; Indonesia, Malaysia, Thailand and Philippines
BOI	Board of Investment
CENTO	Central Treaty Organization
CPI	Consumer Price Index
DAP	Democratic Action Party
DRV	Democratic Republic of Vietnam (North Vietnam)
EPR	Effective Rates of Protection
EOI	Export Oriented Industrialization
EPZ	Economic Processing Zones
FDI	Foreign Direct Investment
FELDA	Federal Land Development Authority
FEER	*Far Eastern Economic Review*
FPE	Financial Public Enterprise
GATT	General Agreement on Tariffs and Trade
GDP	Gross Domestic Product
GNP	Gross National Product
HCMC	Ho Chi Minh City
HDB	Housing Development Board
ICMI	The Association of Muslim Intellectuals
ILO	International Labour Organization
ILSSA	Institute of Labour Science and Social Affairs
IMF	International Monetary Fund
ISI	Import–Substitution Industrialization
JPRS	Joint Press Release Service
LDC	Less Developed Country
MITI	Ministry of International Trade & Industry
MNC	Multinational Corporation
MTUC	Malayan Trade Union Congress

NATO	North Atlantic Treaty Organization
NEP	New Economic Policy (Malaysia)
NESDB	National Economic and Social Development Board
NFPE	Non-Financial Public Enterprise
NGO	Non-Governmental Organization
NIC	Newly Industrializing Country
NIDL	New International Division of Labour (Malaysia)
NPA	New Peoples' Army (Philippines)
NPP	New Population Policy (Singapore)
NSC	National Security Council (US)
NTUC	National Trades Union Congress
OPEC	Organization of Petroleum-Exporting Countries
PAP	Peoples' Action Party (Singapore)
PAS (PMIP)	Parti Islam SeMalaysia (Pan-Malayan Islamic Party)
PDI	Partai Demokrasi Indonesia (Indonesian Democratic Party)
SAL	Structural Adjustment Loans (Thailand)
SATU	Singapore Association of Trade Unions
SDU	Social Development Unit (Singapore)
SEATO	Southeast Asia Treaty Organization
SMI	Small and Medium-sized Industries
SOE	State-owned Enterprises
SSA	Social Structure of Accumulation
STUC	Singapore Trade Union Congress
TNC	Transnational Corporation
UMNO	United Malyas National Organization
UNDP	United Nations Development Program
VGCL	Vietnam General Confederation of Labour

Southeast Asia

Contributors

David Drakakis-Smith

Professor of Geography, Department of Geography, Liverpool University, England

Niels Fold

Assistant Professor, Institute of Geography, University of Copenhagen, Denmark

Jacques Hersh

Professor of Development and International Relations and head of Research Centre on Development and International Relations, Aalborg University, Denmark

Joel S. Kahn

Professor of School of Sociology & Anthropology, La Trobe University, Bundoora, Victoria, Australia

Niels Mulder

An independent anthropologist interested in the cultural dynamics of the contemporary Thai, Javanese, and Tagalog–Filipino societies. He can be reached via P.O. Box 53211, 1007 RE Amsterdam, The Netherlands

Irene Noerlund

Research Fellow, Nordic Institute of Asian Studies, Copenhagen, Denmark

Richard Robison

Professor of Southeast Asian Studies and Director, Asia Research Centre, Murdoch University, Perth, Western Australia

Johannes Dragsbaek Schmidt

Assistant Professor, Research Centre on Development and International Relations, Department of Development and Planning, Aalborg University, Denmark

Somboon Siriprachai

Research Fellow, Department of Economic History, Lund University, Sweden/Thammasat University, Bangkok, Thailand

Arne Wangel

Associate Professor, School of Social Sciences, Universiti Sains Malaysia, Penang, Malaysia/University of Technical Science, Copenhagen

Changing realities of social transition in Southeast Asia

Johannes Dragsbaek Schmidt, Jacques Hersh and Niels Fold

On the brink of a new millennium, the structure of the world capitalist system has taken on new contours as a far-reaching transformation seems to be taking place. Changes in production patterns and the growing importance of financial capital and footloose investments have forced a shift in popular and public response away from an active focus on struggles to resolve contradictions to a more defensive strategy of relying on accommodation and compromise. In addition, the new type of global–domestic interactive social system of accumulation points to a general decreasing influence of states relative to the growing power of international capital.

Furthermore, in connection with globalization, world attention has been drawn to the increasing importance of economic actors and institutions in East and Southeast Asia which are rapidly integrating into as well as challenging the global economy. It should be recalled, however, that this region is not entering the capitalist world system as a previously external area. During the imperialist and colonial era, these countries were component parts of capitalism – although the expression of this relationship was different.

In tandem with relatively high and sustained economic growth, social structures inside these societies are becoming increasingly complex and characterized, to varying degrees, by unequal distribution of material and political rewards. The central questions addressed in this volume seek to explain these emerging structures of economic and social inequality and their concomitant distribution of societal resources. These inquiries intend to refocus scholarly attention away from analysis of an artificial division between state and market in order to identify the agents of social change in the evolution of Southeast Asian societies.[1] A motivation behind this approach is that since the 1980s, the class paradigm has steadily been eroded from development studies. This is, of course, an enigma, especially in the context of this region which, in the 1960s and the first half of the 1970s, was a centre of revolutionary activities where nationalism and socialism based in social movements confronted procapitalist forces and Western powers, principally the United States. Today peace has broken out in Southeast Asia and the future is, albeit in a short-term per-spective, seemingly harmonious and bright.

In contrast to this optimistic picture, historical and comparative analy-ses suggest that much more complex and unbalanced social constellations are emerging in the region. The ebbing of the revolutionary tide accom-panied by a worldwide revival of methodological modifications, ideo-logical deployments and conceptual stretch have outflanked the key concepts, of 'class' and 'social movement', and created a theoretical vac-uum. Two decades of harsh neoliberal and 'post-Marxist' critique have made the class paradigm elastic and vague; the combined political and ideological attack on working classes and trade unionism have made the anticapitalist ethos problematic by putting it on the defensive. To pro-ponents of growth-fetishism, class has lost its credibility as a useful ana-lytical tool and the concept is seen as inadequate to discern and operationalize in any consistent and accepted way the social changes which have taken place in the world in general and in Southeast Asia in particular.

In the current academic world the class concept has undergone radical changes and carries a plurality of meanings. However, it is argued here that it is difficult if not impossible to grasp in any meaningful way the economic and social contradictions between strata and social groupings without reconsidering the concept of class, its adherent theories and their application to analyses of the social context. The expansion of markets and transformation of the processes of production in the emerging economies create wage labour and its opposite component, i.e. the bour-geoisie. That labour too has become a commodity, in Southeast Asia, might be superfluous to point out.

The passing from one form of capitalism with its production pattern and structure to another necessitates a clear meta-theoretical discourse of changes in social relations and linked elements of society. As one of the mentors of Southeast Asian research recently noted, there will be, in the future, a larger place for sociology in studies of the region devoting 'more attention to the urban sector, to labour as well as industry, to the media and modern culture' (McVey 1995: 8).

However, a word of caution is necessary. In order for the notion of class to fulfil its role as an instrument for understanding societal organization and change it has to be comprehended in a dynamic sense. As such there are two ways of looking at the concept of classes.[2] The first of these, which is employed by classical sociology, including different trends within the Marxist tradition, considers class as a structural location, i.e. as a form of stratification, a layer in a hierarchical structure, differentiated according to criteria connected to income, market opportunities or occupation. In opposition to this 'geological model' – as Ellen Meiksins Wood calls it – there is the social–historical conceptualization of class as a social relation between producers and appropriators of surplus labour.

Setting the scene

As mentioned above, one generation ago, Southeast Asia was in revolutionary turmoil. This was not only a continuation of the national liberation movement but also a struggle for some form of socialism to resolve internal contradictions and social conflicts in these essentially agrarian postcolonial societies. Various factors combined to neutralize this revolutionary wave. The US political and military engagement accompanied by the Soviet Union's policy of coexistence and the Chinese Cultural Revolution contributed to this process. At the same time, changes in the *modus operandi* of world capitalism offered American-supported authoritarian (often military) regimes opportunities to implement capitalist growth. The centrality of this evolution and especially its consequences for the societies of the region, together with the attempt to draw some future perspectives, form the substance of several contributions to the present volume. In this respect it is important to focus on the formation of classes and stratifications which have emerged during the process of economic growth in Southeast Asian economies.

Some of these social groups are linked to the boom of export-oriented manufacturing industries, some are connected to an expanding public sector, while others are cashing in on a growing internal market. In this context it is interesting to observe the speed of these changes. With the economic reforms and opening to the outside, new social strata are almost instantly becoming visible in the former socialist countries of Southeast Asia. Notably in Vietnam, centrally planned allocation of resources and socialist ideals of equality are increasingly being replaced by private business investments, market forces and individual aspirations of social mobility. The study of the formation, development, dynamism, and interaction of these new social classes and groups in the still predominantly agricultural economies of Southeast Asia raises a multitude of problems for social research and poses exciting questions.

Of particular importance is the emergence of the 'middle class' and 'working class' as social and political categories. Both are products of economic growth and are in a position to impact on the modernization process from below. What is the size, composition and strength of these new social strata, to what extent are they formed or organized, and to what extent do they identify as classes? Of equal significance is the point that rapid increases in consumption create new social divisions. Since the working class produces more than it consumes, the structure of capitalism requires a corresponding class – one that will consume more than it produces. As production expands and becomes rationalized, it is necessary to increase both the number and diversity of unproductive service workers, which the rising surplus makes economically possible (Urry 1973: 177).

Economic development in the region has undoubtedly resulted in a general increase of national wealth at the aggregate level (as revealed in Table 0.1 and Figures 0.1 and 0.2). However, the emergence of new social groups is evidence that growing inequity has developed simultaneously with the improvement of living standards for relatively large segments of

Table 0.1   General social and economic indicators

	Indonesia	Malaysia	Singapore	Thailand	The Philippines
Population (mid-1992, million)	184.3	18.6	2.8	58.0	64.3
Nominal GDP (1992, US$ billion)	126.4	57.6	46.0	110.3	52.5
Average annual growth rate of GDP (1980–92)	5.7	5.9	6.7	8.2	1.2
GNP per capita (1992, US$)	670	2,790	15,730	1,840	770
Real GDP per capita (1991, US$, PPP adjusted)[1]	2,730	7,400	14,734	5,270	2,440
Consumer expenditure as % of GDP (1992)	50.8	45.5	43.1	54.8	33.6
Share of consumer expenditure on food (1992, %)[2]	53.4	39.1	26.3	33.6	56.5

Sources: Euromonitor plc., *International Marketing Data and Statistics 1994,* The World Bank, *World Development Report 1994,* UNDP *Human Development Report 1994.*
[1] Adjusted by the purchasing power parity factor as defined by the UNDP.
[2] 1991 for Indonesia, Malaysia and Thailand.

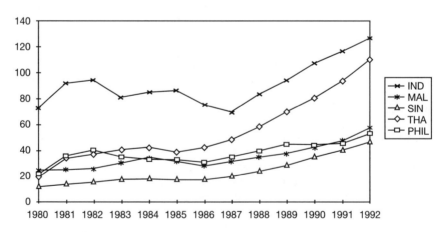

Figure 0.1   The growing economies of ASEAN (US$ billion)
 Sources: The World Bank, *World Development Report* series,
 Euromonitor Plc., *International Marketing Data and Statistics,* various
 issues.

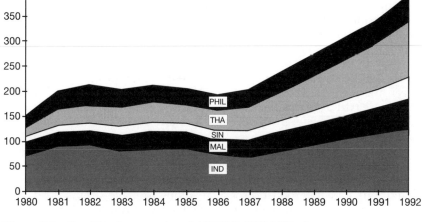

Figure 0.2 Combined economies of ASEAN* (US$ billion)
* Excluding Brunei Darusalam
Sources: The World Bank, *World Development Report* series.
Euromonitor Plc., *International Marketing Data and Statistics,* various
issues.

the population. Thus, recent evidence from the multilateral institutions substantiates that, along with high growth rates, economic inequity in the region has grown considerably during the 1980s and early 1990s, although its scope and nature vary between the different countries. Contrary to conventional wisdom, economic growth, in itself, does not lead to greater equitable distribution. In fact, this gives authoritarian regimes their *raison d'être*, i.e. to control politically the social disparities created by the normal functioning of the market in a capitalist economy.

Left unattended, growing inequalities can eventually lead to social tensions. Therefore, the way quasi-authoritarian states of Southeast Asia deal with increased social differentiation is of crucial importance for the political-economic dynamics of the region. To put it in simplified terms, the choice seems to be between economic growth, inequity and increasing individual consumption on the one hand, and more equitable income distribution and new common social objectives on the other hand. A related issue of particular importance is whether or not social policy is subordinated to the needs for labour flexibility and/or the constraints of international competition.

During the process of economic growth, a whole set of societal issues comes to the fore. As a corollary to the rapid social and industrial transition in Southeast Asia, migration, recruitment and division of labour take new forms. Migration from rural areas to urban centres and immigration or emigration in the region change the ethnic and gender composition of the workforce. This raises the question of whether and how the division of labour is being consolidated according to class stratification. The social transformations in the urban areas of the ASEAN-4 have not

5

only involved the emergence of capitalist and working classes, but the rise of middle classes and of new rich strata; each of these classes being in a process of embryonic segmentation. A most important problem, however, is the relatively large and often poor agricultural population. The incorporation of these people, who form the majority of the population, into the growth process is possibly the most serious challenge facing policy-makers and planners. This aspect represents a significant difference between the East Asian NICs and the ASEAN-4: in the latter case wages have increased at barely half the rate of GNP growth which was about 5 per cent on average from 1965–90 (Asian Development Bank 1992; Morley 1993: 6).[3] The existence of a large reserve army, in the form of a low-paid labour pool, strains class cooperation which might have political consequences in the longer perspective.

At the structural level there are tendencies indicating that traditional institutions and paternalistic social relations are being replaced by the market system based on capital–labour relations. A new management and business elite has emerged which cooperates with foreign companies to various degrees. What support do the power elites or ruling classes get from the emerging middle classes or groups? How are plantation workers, small farmers, and industrial and service workers organized and how do they compare to the middle classes? These are significant aspects in the changing context of these societies.

Whether the emerging groups and new social strata will affect the form and content of popular resistance to authoritarian states is an open question. It is still not evident that the combination of new social and political groups increases popular pressures for democratic political reforms and a more egalitarian distribution of welfare. Establishing the political potential and role of the middle class is a challenge for social analysis.

Understanding social change

There are as many ways of studying changes of social order as there are ways of studying societies. However, in order to understand social transformation it is essential to be comparative. Class as stratification is the first aspect of social structure that needs to be studied. The second crucial aspect is the way the power elite perceives itself, the world, and the way it seeks to legitimate unequal distribution of social resources. The third important element is the global context in which societies exist and change, because this, along with internal factors, has an enormous influence on the direction and nature of the social transformation process (Chirot 1986: 2–4). Taken together, these three elements constitute the investigative core-areas of a conceptual design aimed at studying societies that experience rapid social change.

Achieving growth and industrialization, strengthening the nation-state, eventually raising the welfare of the population, and generally disciplining any late-industrializing society in order to catch up in a relatively short time is, under any circumstances, no easy task. This is related to the workings of the present international system. Future social research may well establish that the 'locus' of accumulation and power did in fact originate in Asia (India and China) before the advent of modern capitalist imperialism established Eurocentric diffusionism (Blaut 1993) as the ideology of modernization. But in the contemporary era of world history it is the latter paradigm of modernity which has become dominant, both in theory and practice. As such it determines the criteria and functioning of the whole as well as the entities within it. The modernization discourse projects present-day advanced liberal and democratic capitalism as the ideal-type model to be emulated by the rest of the world. But in reality, such a projection gives a very distorted picture of a complicated evolution. Thus, when looking at late industrialization and its different social and political aspects, it is not sufficient to look at the mode of functioning of modern Western capitalism. It is necessary to recall that the process of European modernization was itself marked by disharmony and imperialism, as well as by long periods of mass poverty, human displacement and popular resistance to oppression, and that class-based organizations were the dynamic force behind many of the social transformations which took place. Violent internal and international wars paved the way for the early modernization of the capitalist economies. Except for Germany and Japan, the process stretched over centuries and the 'first comers' were not subject to the same pressure of catching up to an already highly industrialized and militarized world. In the past the West was always at the forefront of the advances and did not need to emulate other socio-economic models in order to catch up. Although Karl Marx stated that England was the example to be followed by the rest of Europe in order to industrialize, this could be done on the continent without changing their political systems to conform to the British pattern.

Today, there is a sense of urgency in East and Southeast Asia that allows no time for the luxury of planning in terms of centuries or of passively accepting an inferior or backward status. The clash between established mature societies and 'late developers' attempting to catch up in terms of industrialization and social dynamics is clearly manifest in the unfolding battle of ideas in the world after the dissolution of socialist regimes. These are highlighted in the international media, in the academic community and not least among policy-makers. In 1993, East Asian leaders issued the Bangkok Declaration on human rights, claiming that the 'Asian Way' (harmonious, disciplined, collectivist) is more suited to their societies than the opposite value system of the West (chaotic, unrestrained, anarchic) (*Economist*, 9 July 1994). Such statements ostensibly reflect a clash of civilizations; but this may nevertheless mask a convergence of political ideologies as part of a new transnational offensive against social reforms as well as against regulation of labour and capital markets in both the East and the West.[4] In the West, the competition from Asian economies, based on cheaper labour, is used to discipline and instil fear of job losses, if profitability of enterprises is not guaranteed. In short, these discourses are presented by civilizationists as a conservative counter-offensive in the debate about class struggle and social contradictions in society. Different norms, values, civilizational attitudes, and political cultures based on gender, religion, race, ethnicity and nationalism, together with post-modern attacks on class, serve to deconstruct the concept of class by multiplying the dimensions of inequality and stratification, thereby articulating a blurred theoretical discourse (Lieberson in Grusky 1994: 22).

This is also partly reflected in the existing compartmentalization of the literature on Southeast Asia where there is a tendency to look at the region either from an economic and regime-type perspective or from a predominantly culturalist angle. Without entirely abandoning these paradigms, the present volume attempts to 'bring classes back in'.

It is possible to identify four major approaches to the study of Southeast Asian capitalist development. The neo-modernization perspective reverses Max Weber's Eurocentrism and the notion of the Protestant ethic by pointing to Confucianism as a positive factor;[5] the economistic approach favours the 'magic of the market' explanation and points to free trade and free markets as the main determinants of growth and welfare. The statist view considers a strong and activist political entity as the most decisive element in explaining how wealth is created and distributed. World-system and dependency theorists look at historical and geopolitical relations of societies to find evidence to explain why only a few selected countries located in East Asia, all of them neighbours to Communist states, were successful in becoming US-initiated 'capitalist showcases' (McVey 1992: 9; So and Chiu 1995).

The culturalist explanation, not surprisingly, has its primary proponents in East Asia; but this type of neo-modernization theory has also gained popularity among some Western academics (Berger and Hsiao 1988).[6] The market approach is not only favoured by mainstream thinking in multilateral organizations such as the World Bank and IMF, but is also a component of the so-called 'Washington Consensus', where neoclassical, neoinstitutionalist, and rational choice approaches combine to form the dominant paradigm (Chen 1979 and 1989; World Bank 1993).[7] But just as importantly this latter discourse also has its supporters in the domestic contexts of Southeast Asian countries among pragmatic technocrats who, in opposition to nationalist segments of the bureaucracy and bourgeoisie, act as the local agents of transnational interests trying to transform the state into a deregulated transmission-belt between international and national capital.

Recently however, the statist argument has gained considerable strength, especially in small, isolated but increasingly powerful milieus among new strategic traders and academics of the Anglo-Saxon world, where the results of two decades of neoliberal onslaught have created a gridlock of rising polarization, increasing inequalities and poverty. For its proponents the construct of the statist approach derives from such concepts as the capitalist developmental state, governed economies in East Asia, and the Japanese model. While it is particularly strong in England and the United States, it has gained influence in Southeast Asia as well.

Scope, arguments and theoretical explanations

The principal objective of the present book is not to criticize these approaches per se but rather to (re)introduce a more society-based focus on classes and emerging inequities in Southeast Asia. Radical transformations in social and economic structures in the region have led to new conceptualizations of group interests and political values as well as societal pressures. In the wake of the economic growth of the region, class tensions were pacified and redefined as different leading groups, taking advantage of the opportunities, rewrote the political, social and economic rules to accommodate their interests at the expense of the disadvantaged and marginalized segments of society. We feel that the comprehension of development in Southeast Asia has been badly served by existing explanations. In order to achieve a higher degree of clarity and common sense, we propose that the East Asian and Southeast Asian experience can be better understood by awareness of three sets of problems: the kind of social contract which might or might not emerge in the region; the correlation between political control and social organization and specifically its political manifestations; the impact of the international context on class segmentation; and new conflict patterns.

Is it possible to provide one stylized theoretical explanation for these complex phenomena which also touches upon philosophical aspects? Are they the outcome of the concrete agency of social forces or do they mirror the model of the East Asian authoritarian *nomenklaturas* only? Are the social compacts in Southeast Asia to be understood as negotiated agreements that remove sources of potentially destabilizing conflicts from the political agenda or the outcome of pragmatic and non-ideological state policies? Is the problem of order in these societies to be solved through a Hobbesian solution where classes and strata submit to the authority of the Leviathan, i.e. the state, in the absence of which there will be chaos? Or can order versus disorder be defined through a Lockean conceptualization of natural rights to life, liberty and property as being best protected by the state? Or is it impossible without implementing Rousseau's ideal-type society where achieving equality is part of a process which can never be realized? The solution according to this line of thought is direct democracy, expressed through the 'general will' which affords the greatest protection for the individual.

In this connection it is important to keep in mind that in the late catching-up processes of East and Southeast Asia, the state is fetishized as an autonomous entity. But the state in these societies – as is the case elsewhere – should be understood as a social form, that is a form of social relations, instead of conceptualizing it in a rigidified form (Burnham 1994: 226).

Does the East Asian model represent a challenge to conventional wisdom to the extent that while liberalism loudly supports the celebration of the market, uncontrolled international competition, wealth creation, anti-socialism, and anti-unionism, behind closed doors it endorses state intervention, authoritarianism, anti-market, and political control? What are the consequences of this seeming paradox in ideology and especially in its political expression? This question leads automatically to the problem of whether analysis can discern an independent or unique East Asian model or whether Westernization is unavoidable. Is Western rationality of a 'modern' social order an absolute frame of reference in the East Asian context? In this relation it is essential to distinguish, as Susan Strange reminds us, that: 'Most actual national systems in any case are compromises with both political liberalism and economic liberalism'. The former refers to democracy with freedom of expression, participation in policy-making. Economic liberalism entails responding to supply and demand, not at administered or manipulated prices, but market-determined (Strange 1994: 216). In other words, it is quite possible to respect market mechanisms in an environment of political illiberalism.

International pressures on Southeast Asian societies come in many forms and constitute a third problem dealt with in the first two chapters. After the colonial period, the area – as mentioned previously – was submitted to the interference of external forces affecting the social evolution of these formations. Besides the historical dimension, the economies of these regions have become enmeshed in the international division of labour, thus adapting their socio-economic strategies to the demands of the world market. This has of course had internal consequences for these societies. It is often claimed that Southeast Asian capitalism has been successful in containing protests against inequity in the distribution of welfare and in offering opportunities for stratified advancement. In another view, the highly unpopular externally induced macroeconomic adjustments which have not been equally shared are seen as an expedient formality and only a matter of getting the infamous trickling down effect to work. The question emerges as to how the different collective actors or components of the social structures react to these highly diversified effects of the changing global context.

Bringing back the notion of class can conceal more than it reveals. As a way of ordering the theoretical argument, it may be helpful to follow David Slater's point about the four main problems related to the centrality of class analysis (1992: 295). Clarity of concepts and frames of reference need to be reinstated in order to understand its analytical operationality.

1. Class, as an abstract concept or mental construct, is not capable of social action or agency; just as *gender* does not act, but women and men do. Equally, in contrast to a political party, trade union, state agency or firm, etc., classes as social aggregates do not have any means of taking decisions or acting upon them.
2. It is assumed that classes have 'objective interests', which result from the overall structure of class relations, and which then function, *a priori*, as the necessary basis for the mobilization of actors, divorced from any analysis of the varied constitution of the social subject or the dynamic of collective wills. The failure of classes to become conscious of their 'interests' gives rise to the notion of 'false consciousness'.
3. It is further presupposed that a class in struggle, for instance the proletariat, has a consciousness that emanates from or is centred at the

point of production. In this way, a fixed foundation is assigned a pre-given significance in the analysis and interpretation of consciousness.
4. Lastly, the proletariat, or more broadly defined, the working class, has been conceptualized as *the* privileged revolutionary social subject, thus dispensing with the need to explain the processes whereby varying forms of political subjectivity are constituted, and through which the propensity to act collectively may emerge under certain specific circumstances.

In the Marxist tradition, the concept of class has been used mainly in the context of analyses of social conflict and change, with the labels 'civil society' and 'new social movements' having gained popularity and marking a predominantly Western perspective evolving out of a critical evaluation of orthodox Marxism. Since the mid-1970s some of these new theories have been branded as post-Marxist, because of their rejection of both economic reductionism and a pro-proletarian stance. These theorists stress the novelty of the 'new social movements' and contrast them with the 'old social movements', mainly of the socialist type. More importantly, they locate the new movements within the broader realm of 'civil society' rather than within the sphere of property relations. They also embrace a new form of partisanship which is rooted in communicative ethics and which abandons the revolutionary eschatology of the Old and New Left.[8]

In the Weberian tradition, class serves more as a *categorical* concept referring to market-generated patterns of inequality in opportunities in life (differentiated access to goods and services). Neo-Weberians reject the polarity of classes, emphasizing the importance of authority relations and mobility in social formation and contrasting class conflicts with status group and party-political divisions (Pakulski 1993: 134).

In this introduction, class is understood as a real historical relationship and not merely as a subjective analytical category in the mind of the analyst. Whether or not social classes exist is a matter of investigation. The social basis for the existence of classes comes from the way in which people are positioned in production processes, but though production creates the potential for classes, it does not make classes. Social practices shaped by events give people the common experience of class identity and of collective action. Some of these practices are oriented toward production, such as trade unionism, some toward the state, in the form of political parties and movements. However, such processes do not operate in a political vacuum. Powerful forces are at work, supported by the dominant elites, that obstruct class identification and formation among emerging subordinate social groups. This can take the form of occupational fragmentation, exploitation of ethnic and religious identities and symbols, creation of helplessness in the face of the coercive repression of states and dominant groups (both official police and military repression and unofficial death squads), spreading of consumerism, and petty-bourgeois aspiration to individual upward mobility (Cox 1987: 355 and 389–90).

In Third World countries that have succeeded in achieving extended periods of high economic results, the state depends on its control over official trade unions to ensure the supply of trainable, docile, and cheap labour on which growth is based. To the extent that it can prevent the emergence of an alternative labour leadership, state corporatism underwrites the profitability of industry. But generally it has not been able either to gain the complete allegiance of workers or to suppress entirely the expression of worker interests and discontent. This is especially so where there has been some experience of autonomous labour organization. An official union is generally regarded by workers as an arm of the state or of industry, not as a worker organization (Cox 1987: 384–5). This is also the case in Southeast Asia; but alternative unions outside the jurisdiction of the state are emerging, especially in the Philippines and Malaysia, but lately also in Thailand and Indonesia.

One of the hallmarks of this class-based approach is its focus on the way processes of historical change are shaped by the patterns of conflict and alliance between classes or factions. Recent work on political development provides a good example. Long-standing cross-national findings on the positive relationship between democracy and growth have been interpreted in an analytical class framework. In this view, economic development is associated with political democracy, not because of some amorphous connection between a more differentiated occupational structure and a penchant for parliamentarianism, but as a result of historical patterns of conflict and alliance between different classes. This analysis differs from its more orthodox predecessors of dependency and Marxist theory by putting emphasis on the ability of abused classes to influence historical outcomes through class action. But, it also emphasizes the variability of class alliances and the importance of legacies of past associations for current outcomes (Evans and Stephens 1988/89: 720–1).

Can contemporary class analysis draw anything of value from the historical contributions of the classic theoreticians? In the case of Karl Marx, what has been retained is essentially the 'political class analysis'. Class defined by economic position remains central but the impact of class membership on political behaviour is assumed to be contingent on a variety of social-structural and historical factors. Important 'grand theoretical' interpretations of Max Weber have also been rejected. 'Value determinist' readings of Weber, in which shifts in the internal logic of cultural systems emerge independently of change in social structures, thereby providing the key stimulus to developmental progress, are clearly not compatible with a comparative political economy approach. Equally alien is the reading of Weber that defines development in terms of a teleological impetus toward ever more pervasive rationality. What is retained is the historical and institutional side of Weber with its sensitivity to the way rational, social and organizational structures are reshaped by considerations of power and conflicting interests, and Weberian careful portrayal of the interaction of cultural and social structural features. In addition what is worth clinging to is the conviction that history must be written internationally and that the national developmental trajectories must always be understood as shaped by the context of the international system (Evans and Stephens 1988/89: 728). This present book situates the trajectories of social change in Southeast Asia in their contemporary as well as historical and international contexts with a renewed focus on a societal perspective.

Suggested below is a theoretical framework for analysis of class inequity. An important point of departure is the relationship between state policy-making, strong economic growth and the uneven spread of benefits. In addition it is difficult to explain why some social groups receive more income than others and why some classes emerge much faster than others if the political correlates between internal and external politics are not taken into consideration simultaneously.

The following aspects of societal evolution complement the above theoretical considerations:

1. The construction, guidance and direction of emerging classes are partly the outcome of implemented state policies and partly depend on market allocations and the interplay between the market and the state.
2. It is the redistributional consequences of economic policies and the state's choice of development strategy which partly determines winners and losers, i.e. who gets what, when and how. The balance of strength between social groups is an important determinant of the type of regimes needed.
3. To use a Scandinavian expression, not only does the state 'create a snake at its own breast' by implicitly supporting an up-coming middle class, but this class soon makes its own demands on the state by exerting tremendous pressures for collective consumption, i.e. public goods and social services. This process is mainly manifested through political expressions but takes place in a context of growing inequality and marginalization of the majority of the population.
4. No country in Southeast Asia, except perhaps Malaysia, Singapore and Vietnam, has in relative terms a binding and functioning social contract in the Rousseau mould. The legitimacy of the governments and state bureaucracies in Thailand and Indonesia, and to a certain extent in the Philippines, is based on economic growth and an attempt to promote social order. What is quite unique in Malaysia is the fact that the Bumiputra content of the New Economic Policy and Vision 2020 can be interpreted as a sort of social contract. The Bumiputra policy, successful or not, provides a strong incentive to force the Malay population into a social compact with the state apparatus. Some scholars suggest that the royal institution in Thailand and the Pancasila ideology in Indonesia perform a similar function. But these models are extremely fragile because they are personalized and therefore laden with potential uncertainty. To draw a comparison with the Philippines during the Marcos regime – which, by the way, is very similar to the situation of countries in Latin America – it is clear that when the legitimacy of the state, economic growth and a personalized social contract break down, the result is chaos, instability and lack of popular trust in even the basic functions of the state and the performance of the regime.

Structure and content of this volume

The collection of essays assembled here represents a diversity of approaches and theoretical points of departure. As editors we have not tried to offer a common methodology. However, the intention has been to show how class formation develops, and to identify patterns of class transformation and inequalities in Southeast Asia by means of placing social contradictions and social conflicts in the context of capitalist growth.

The first part of this volume presents structuralist perspectives that attempt to explain the dynamics and constraints of the emerging class structures in Southeast Asia. The first chapter emphasizes the role of the international environment, while the second contribution focuses on the linkage between state policies and social change. In contrast, the third chapter examines the rapid economic development in Southeast Asia as driven by strong local capitalist forces and the middle classes.

Jacques Hersh uses an international structuralist perspective to situate these societies. The strength of the argumentation lies in the emphasis placed on the historical context and the concomitant link between US geopolitical and geoeconomic interests in creating functioning and successful capitalist alternatives to the former Communist threat from China and Indochina. According to this line of thinking, the East Asian model with its (in)famous middle class is predominantly the result of favourable external conditions: first the political–military and economic strategy of the United States and secondly the Japanese economic expansion in the area. But in the past decade or so class constellations and contradictions have developed according to an internal logic, which exclusive focus on the external influences may obscure.

Johannes Dragsbaek Schmidt introduces a combined analytical construct between the domestic and the global context, with emphasis on the role of the state as a transmission belt between the internal and external constraints and possibilities of social change and welfare. Accordingly the state is seen as much more than a technocratic entity as it fulfils its role of social control during the capitalist growth process. It is pointed out that in Southeast Asia, there is a general trend towards relinking growth and political participation which potentially might give rise to social tensions. Schmidt furthermore discusses the influence of multilateral institutions and TNCs on shaping the technocratic bias toward the export-oriented type of economic strategy and societal organization.

Richard Robison presents a revisionist but structuralist analysis which derives its theoretical origin from Ralph Miliband and James Kurth. The argumentation is based on the proposition that politics and political outcomes are determined by social structural factors (Hewison et al. 1993: 4). His present inquiry analyses the rise of the middle classes in Southeast Asian countries with special focus on Indonesia. According to Western and liberal orthodoxies, the emerging bourgeoisie and middle classes are seen as bearers of modernity and rational culture. Several questions are raised with regard to situating the middle class, new and old, as an intermediary between the working class and the bourgeoisie; Robison concludes that this particular stratum, i.e. the middle class, has had an important impact on political contradictions and conflicts in the societies of the region. The fact that the middle class is a vast and internally differentiated social category complicates a clear definition. Robison discusses the space given by the political system for middle-class politics and the dependence of the middle class on the state for delivering the goods. At a time when lower-middle-class youth are becoming frustrated with the political system, the state is thus weakened.

The second part of this book introduces the problem of the role of the state in influencing the autonomy of the cultural sphere by way of focusing on classes, culture and political legitimation in Malaysia, the Philippines and Singapore.

Joel Kahn and Niels Mulder as representatives of a culturalist and post-modernist perspective, respectively, point to a simultaneous process of accelerating disintegration and dissolution characterized by the loosening of social and political bonds; the delinking of mass support to political parties and social movements; the erosion and levelling of classes, lifestyles and established communities; the end of ideologies and a general end to the narratives of modernity (Betz 1992: 93–4). Joel Kahn's analysis is critical of the dependency and Marxist constructs of the middle strata. With empirical focus on the middle classes and the role of the intellectuals in Malaysia he develops an approach centred on the cultural habits and symbols of these emerging strata. The neo-modernization of Malaysian society is seen as the result of 'party capitalism' rather than a process of state capitalism. Because of the ethnic dimension, the fact that the policies of the dominant party in government favour the Malay population affects the composition of the social classes. What we witness as a result of this strategy is an 'Islamization of the middle class or, what amounts to the same thing, the middle-classization of Islam.' This, in turn, strengthens the legitimacy of the government.

In his chapter, Niels Mulder follows in this direction with a focus on the rise of the middle class and its room for manoeuvre in the public sphere in the Philippines. His emphasis is placed on defining middle class membership culturally in the urban context. The educated strata are seen as having betrayed the nationalist and social engagement of the earlier generation. The drive to career-orientation has created a politically indifferent and socially unaware student body. This is related to the American success in having promoted the superiority of US culture and historical interpretation of the history of the Philippines. The process is seen as leading to an acculturation of the country which nationalism has not been able to overcome. The result is paradoxical: what appears as a developed civil society is not so much pressing its demands on the state and the political system as turning its back on it.

Focusing on Singapore, David Drakakis-Smith addresses the question of why the middle class is a construction of state strategy based on ethnic preferential policies. His approach is based on a combination of state theory and ethnicity which includes the question of gender as well. While ethnicity has been largely ignored by development economists (even though independence and uneven development accelerate conflicts between ethnic groups in multi-ethnic societies) it is important to bear in mind, as David Drakakis-Smith points out, that ethnic labels often conceal and obscure class cleavages. The social engineering of the state in Singapore aims to install a national ethos based on economic rationality and a meritocracy in this multi-ethnic society. This leads to 'human management' and the attempt to influence even the most intimate spheres of individual life. The strategy for creating a better-educated technocratic stratum based on human resource management by the state similarly begets a more disadvantaged or less educated underclass (local or migrant) which finds itself in persistent deprivation. This is seen as the contradiction which is likely to destabilize this nation-state project.

The third part of this volume revolves around the problems of macroeconomic strategies involved in the control of labour and the extraction of a surplus from industrial workers. Capital accumulation based on export-oriented industrial strategies demands institutionalization of regulations and control at both the micro and macro levels of economic activities. The importance of the nexus between the state, the market and the work force cannot be underestimated. Nor should the contradiction between industry and agriculture be ignored, as made clear in the last contribution by Somboon Siriprachai.

Although the ideological shift against state intervention in the developing countries has been in vogue for some years, the French regulation school has accepted the difficult challenge of presenting an alternative theoretical construction for the understanding of social institutions and their function, namely that of managing the conflict over distributional benefits. Building on this framework, Niels Fold and Arne Wangel offer an analysis of shifting patterns of labour regulation in Malaysia. Keeping wages low has been a primordial concern of the economic and political strategy of the Malaysian state. The authors examine the various methods and measures which have been utilized. The structural weakness is also highlighted. Thus the shift to a growth regime based upon productivity increase is confronted with absolute skill shortages within the work force without labour being able to tilt wage formation to its advantage. The systemic rationality of these forms of labour regulation in the prevailing growth regime is explained by the lack of potential for a general increase of labour productivity in the manufacturing industries. Despite growing differentiation within the work force, the hegemonic leadership manages to contain any social contradictions that might develop in this process.

In her contribution, Irene Noerlund provides a detailed analysis, based on field research, of the role of new labour laws and institutions in the emerging capitalist economy of Vietnam. The transition from one labour regime characterized by socialist industrial relations to a capitalist-type labour market is a process which, after initial approval, is meeting increasing resistance among workers in the private and foreign-owned enterprises. In the public sector, workers are still subject to the former system of industrial relations. Privatization has not reached the large state-owned industries. In the context of market reforms in socialist countries, the Vietnamese seem to be following the Chinese strategy of gradual liberalization of the economy while trying to maintain political control by the Communist Party, rather than the process in the Soviet Union where political liberalization preceded the market reforms.

Somboon Siriprachai's chapter diverges from the other contributions by placing its point of departure in a neoinstitutionalist theoretical framework. Somboon's contribution deals with the role of the state, industrialization and the emerging social inequalities in Thailand. Seen from this perspective the author argues that Thailand failed economically because of what he terms the dominance of laissez-faire in the state's economic policy-making. According to this position Thailand ought to have engaged in a process of more active learning from the experience of three East Asian NICs (Japan, South Korea and Taiwan). This contradicts a widely held view, as others point out that Malaysia, Indonesia, Thailand and even the Philippines, during the Marcos regime, did in fact emulate important elements of the Japanese model as defined by Chalmers Johnson (1982; 1986), but failed in several important measures such as land reform and welfare (Schmidt, forthcoming). This last aspect is made clear by Somboon who sees the rural question as a basic weakness in the Thai strategy. The tribute which the agricultural sector is paying to industrialization and urbanization has negative consequences for the majority of the population.

Before concluding this introduction, it ought to be repeated that the editors' intention has been to introduce different approaches, methods and typologies as well as case studies related to the study of emerging classes and growing inequalities in Southeast Asia. The theme is intended to be understood in a multidisciplinary sense, involving academic disciplines such as sociology, anthropology, political science, economics, political economy, social and economic history, human geography and communication science.

In essence, and this cannot be emphasized enough, the approach reflects a shift from explaining high growth to looking at its outcome. In this connection it is important to realize that the sustainability of the economic development models in Southeast Asia in the near future will depend on the social contracts which will determine change as well as participation in the world economy. Related to this question is the role of the working classes and trade unions, i.e. not principally that of the middle class as numerous scholars misleadingly suggest. The World Bank, IMF and other multilaterals are not only afraid of worker-based social change, but together with national governments in East Asia, these institutions make deliberate efforts to avoid the question of the increasing importance of these actors. This is reflected in new social configurations and rearrangements of social order in a region experiencing rapid change.

Of course not all the contributors to this book agree with the preceding discussion. This is well reflected in the heterogeneity of what we know and choice of theoretical position as well as research tradition and methodology. Notwithstanding these differences, most authors have managed to address the important questions related to the emerging class constellations and inequalities in Southeast Asia.

1.   In this book Southeast Asia is restricted to Thailand, Malaysia, Indonesia, the Philippines (also referred to as the ASEAN-4), Singapore and Vietnam. However, the increasing importance of the rest of geographically and politically defined Southeast Asia – Brunei, Burma, Laos, and Cambodia – to the rest of the region might well have been an argument for including these countries. They have been omitted from this study simply because the first is an exception as a single small oil-producing city-state while the three others share similar problems with others of World Bank LDC status. On the other hand the concept of East Asia denotes Japan, the so-called Newly Industrializing Countries (NICs – South Korea, Hong Kong and Taiwan), and China.

2.   This passage leans heavily on Meiksins Wood (1995: chapter 3).

3.   Average GDP growth rates for the ASEAN-4 have been above the average of other developing regions. In the period 1971–80 the average was 7.4 per cent, with an inflation rate of 13.6 per cent. 1981–90 saw a growth rate of 5.2 per cent and an inflation rate falling to 7.5 per cent. Population growth rates have declined from an average of 2.6 per cent in the period 1965–80 to 2.3 per cent in the period 1980–87.

4.   See Huntington (1993) and for an interesting critique, see Rodan (1995).

5.   See the discussion about the applicability of Weberian and neo-Weberian theory to studies on Southeast Asia and further references in Wim F. Wertheim (1995: 17–29)

6.   See the critical discussion and references in the introductory chapter in Taylor and Turton (1988).

7.   See the discussion in Henderson and Appelbaum (1992: 13–15).

8.   For an application of a similar kind of theorizing to the Southeast Asian context, see Hewison and Rodan (1994: 235–63)

Bibliography

Asian Development Bank (1992) *Asian Development Outlook*, Manila

Berger, Peter L. and **Hsin-Huang Michael Hsiao** (eds) (1988) *In Search of an East Asian Development Model*, Transaction Books, New Brunswick, N. J.

Betz, Hans-Georg (1992) 'Postmodernism and the New Middle Class', *Theory, Culture & Society*, **9**(2)

Blaut, J. M. (1993) *The Colonizer's Model of the World. Geographical Diffusionism and Eurocentric History*, The Guilford Press, New York

Burnham, Peter (1994) 'Open Marxism and Vulgar International Political Economy' *Review of International Political Economy*, **1**(2)

Chen, Edward K. Y. (1979) *Hyper-growth in Asian Economies: A Comparative Study of Hong Kong, Japan, Korea, Singapore and Taiwan*, Macmillan, London

Chen, Edward K. Y. (1989) 'East and Southeast Asia in the World Economy: Issues, Problems and Prospects', *Copenhagen Papers on East and Southeast Asian Studies* (4)

Chirot, Daniel (1986) *Social Change in the Modern Era*, Harcourt Brace Jovanovich, Inc., Orlando, Florida,

Cox, Robert W. (1987) *Production, Power and World Order. Social Forces in the Making of History*, Columbia University Press, New York

Evans, Peter and **John D. Stephens** (1988/1989) 'Studying Development Since the Sixties. The Emergence of a New Political Economy', *Theory and Society*. Special Issue on Breaking Boundaries: Social Theory and the Sixties, **17**(5)

Grusky, David B. (ed.) (1994) *Social Stratification. Class, Race, and Gender in Sociological Perspective*, Westview Press, Boulder, Colorado

Henderson, Jeffrey and **Richard P. Appelbaum** (1992) 'Situating the State in the East Asian Development Process', in Richard P. Appelbaum and Jeffrey Henderson (eds) *States and Economic Development in the Pacific Rim*, Sage, Newbury Park, California

Hersh, Jacques and **Johannes Dragsbaek Schmidt** (eds) (1996) The Aftermath of 'Real, Existing Socialism' in East Europe. *Between Western Europe and East Asia*, Macmillan and St Martin's Press, London and New York

Hewison, Kewin and **Garry Rodan** (1994) 'The Decline of the Left in South East Asia', *Socialist Register* 1994, The Merlin Press, London

Hewison, Kewin, Garry Rodan and **Richard Robison** (1993) 'Introduction: Changing Forms of State Power in Southeast Asia', in Kewin Hewison, Richard Robison and Garry Rodan (eds) *Southeast Asia in the 1990s. Authoritarianism, Democracy and Capitalism*, Allen & Unwin, Sydney, Australia

Huntington, Samuel P. (1994) 'The Clash of Civilizations?', *Foreign Affairs*, **72**(3)

Johnson, Chalmers (1982) *MITI and the Japanese Miracle*, Stanford University Press, California

Johnson, Chalmers (1986) 'The Nonsocialist NICs: East Asia', *International Organization*, **40**(2), Spring

Lieberson, Stanley (1994) 'Understanding Ascriptive Stratification: Some Issues and Principles', in David B. Grusky (ed.) *Social Stratification: Class, Race, and Gender in Sociological Perspective*, Westview Press, Boulder, Colorado

McVey, Ruth (1995) 'Change and Continuity in Southeast Asian Studies', *Journal of Southeast Asian Studies*, **26**(1) March

McVey, Ruth (1992) 'The Materialization of the Southeast Asian Entrepreneur', in Ruth McVey (ed.) *Southeast Asian Capitalists*, Cornell Southeast Asia Program, Ithaca, New York

Morley, James W. (ed.) (1993) *Driven by Growth. Political Change in the Asia-Pacific Region*, Studies of the East Asian Institute, Columbia University, East Gate, M. E. Sharpe, New York

Pakulski, Jan (1993) 'Mass Social Movements and Social Class', *International Sociology*, **8**(2), June

Rodan, Garry (1995) *Ideological Convergences Across 'East' and 'West': The New Conservative Offensive*, Development Research Series, Working Paper No. 41, Dept. of Development and Planning, Aalborg University

Schmidt, Johannes Dragsbaek (forthcoming) 'The Challenge From Southeast Asia: Between Equity and Growth', in Chris Dixon and David Drakakis-Smith, *Uneven Development in South East Asia*, Avebury, Aldershot

Slater, David (1992) 'Theories of Development and Politics of the Post-modern – Exploring a Border Zone', *Development and Change*, **23**(3)

So, Alvin Y. and **Stephen W. K. Chiu** (1995) *East Asia and the World Economy*, Sage Publications, Thousand Oaks, California

Strange, Susan (1994) 'Wake up, Krasner! The World has Changed', *Review of International Political Economy*, **1**(2)

Taylor, John G. and **Andrew Turton** (1988) 'Introduction', in John G. Taylor and Andrew Turton (eds) *Sociology of Developing Societies. Southeast Asia*, Macmillan, London

Urry, John (1973) 'Towards a Structural Theory of the Middle Class', *Acta Sociologica*, **16**(3)

Wertheim, Wim (1995) 'The Contribution of Weberian Sociology to Studies of Southeast Asia', *Journal of Southeast Asian Studies*, **26**(1) March

World Bank (1993) *The East Asian Miracle. Economic Growth and Public Policy*, Oxford University Press, New York

Wood, Ellen Meiksins (1995) *Democracy Against Capitalism*, Cambridge University Press, Cambridge

Part 1 The internal and external dynamics of social conflict in Southeast Asia

Asia is not going to be civilised after the methods of the West. There is too much Asia and she is too old.

Rudyard Kipling

CHAPTER 1

The impact of US strategy: making Southeast Asia safe for capitalism

Jacques Hersh

The thrust of East Asia on the world scene has caught the attention of various sectors of the international academic community as well as decision-makers in the world's economic and political centres. Although predictions of European decline relative to Asia were made in the past, present-day claims, according to which the next century will see a shift of the core of world capitalism from the Atlantic to the Pacific, have become actualized by the economic growth of the East Asian area, first in Japan, then in the 'Four Tigers' (South Korea, Taiwan, Singapore and Hong Kong) and then among the ASEAN countries. Last but not least, China has entered the race of economic growth and is being followed by Vietnam.

Although there are national and regional differences between the countries, they are part of the same dynamic process. Seen from a strategic angle, both geoeconomic and geopolitical, there is some justification in using the convenient label of East Asia – without, though, forgetting past and present distinctions between the entities – when discussing Asian capitalism and US engagement in the area since the Second World War. It is interesting to note, as far as the conceptualization of the Southeast Asian region is concerned, that there was a 'coincidence between Southeast Asia's birth as a concept and the triumph of American world power' (McVey 1995).

The Asian dynamism the world has witnessed raises some new questions and problems. How is this 'coming of age' of Asia, to use a paternalistic and Eurocentrist expression, to be interpreted? Why now and not before? In the nineteenth century there was a European tendency to base explanations for the 'backwardness' of these societies on racial theses. With the development of new social sciences, the causality of this state of affairs was ascribed to the relationship of ideology and social structures. Understanding became focused on a culturalistic approach whereby a comparison between Christian puritanism and Confucian values was related to the process of capitalism. On the basis of such findings Max Weber expressed scepticism concerning the prospects of capitalist development in the Confucian cultural sphere (Weber 1968).

In more contemporary times, it may be recalled that not many years ago, the question which was asked about East Asia was: why so much political turbulence and why so little economic growth? The enigma is further accentuated when considering the fact that other regions of the world were perhaps better or at least just as well endowed from the hand of Mother Nature.

In this connection, it is worth noting that, at the time of decolonization, there were no great differences in the economic levels of East Asian and African former colonies. In the 1960s, the World Bank informed us, Ghana and South Korea had the same GNP per capita (approximately $230). Today South Korea is 12 times as prosperous as Ghana (World Bank 1991: 268–9, 352–3). At the point of departure both countries were agrarian economies with a common history of colonial rule. Of course Japanese and British colonialism had shown differences in their modes of functioning, with different impacts on the respective colonial societies. Japan had developed infrastructure and some industries, mostly in northern Korea, as well as establishing a regional division of labour, including Taiwan. Nonetheless, both Ghana and South Korea were faced with great difficulties in trying to transform their postcolonial economies. While the present-day culturalist approach sees an advantage in Korea's historico-cultural coherence – in contradiction to the Weberian thesis on Confucianism – economists can point to Ghana's more promising prospects because of the country's more abundant natural resources. Furthermore, the Korean War had left the peninsula devastated and reduced future prospects by cementing the division of the nation.

Levels of analysis

From a developmental theoretical perspective the task is to explain the different results and trajectories taken by the various countries in what came to be known as the Third World. When examining the phenomenon of outstanding economic growth which the East Asian region has attained it is necessary to ascertain the appropriate level of analysis.[1] Socio-economic outcomes can be looked at from various perspectives. We can choose to focus on the role of the individual, on social institutions or on the international environment. Some of the classic analytical variables of changing reality are based on the following aspects: the importance of culture, the functioning of markets, the significance of political institutions, the distribution of economic and political power between groups and classes concurring in the creation of the economic surplus and last but not least the interplay of countries within the interstate system. Other weighty elements which both influence a dynamic process of transformation and are affected by it can be added: migration, urbanization, social organization, ethnicity, family and gender.

In the study of development or transformation processes it is necessary to explore social outcomes based on the behaviour of collectivities. The reason is that individuals do not live and act in isolation but do so in the context of societal organizations, i.e. collectivities. The task that arises for social sciences, then, is to determine the most appropriate way to understand the mode of operation and behaviour of collectivities.

In the Anglo-American academic tradition, the focus of the scientific method starts at the level of the individual and then moves on to the larger unit. According to this approach it is only after understanding the attributes of individuals – rationality, culture and emotion – that the type and behaviour of societies can be determined. In other words, the micro-level of analysis takes precedence over the macro analysis. Emphasizing human nature as a determinant for the need to create a Leviathan to, in turn, control people's destructive tendencies belongs to the realm of Hobbesian political philosophy.

Another tradition reverses the mode of reasoning by putting emphasis directly on the macro-level of analysis. Systems are thus considered as primordial to individuals, to a fundamental degree, exhibiting identity and behaviour imposed upon them by the structure. Consequently, macro conceptualization is given precedence over the micro-level. This approach, which is linked to the sociology of Emile Durkheim, differs from the Anglo-American mode of analysis in its point of departure. However, this 'structuralist' way of looking at collectivities and individuals tends to reduce the possibility of social change originating at the grass roots.

In contrast to these two traditions, Marxism attempts to surmount the contradiction. This is done by considering the duality of culture and structure as a dialectical relationship whereby individuals, though products of society, are nevertheless capable of action to transform it (Marx 1958: 246).

To serve our purpose, this discussion has to be brought up to a higher plane. Just as individuals can shape and be moulded by the collectivities in which they live, these units exert and are submitted to similar mechanisms in the global system. In other words, the behaviour of states influences the international structure while simultaneously being exposed to the impact of the organization of the world political economy. We know of course that not all individuals are equal in collectivities. Likewise in the world system some nations are ranked higher than others and exert greater influence and power.

To continue along the line of determining the unit of analysis, it is suggested here that, in order to understand the dynamism of East Asia since the end of the Pacific War – as the Second World War is known in Asia – and more specifically since the 1960s, the external dimension has to be brought in. That is the role of states and the weight of the world system. The way the international environment opens possibilities and forecloses others has an impact at the level of national politics. Without considering the post-World War II conditions, the leadership position and strategy of the United States in the international political economy, it is difficult to comprehend the specific development of East Asia. That does not of course mean reducing the significance of individuals and collectivities in societal change, but these have to be seen in a dialectical relationship to the opportunities and constraints which the international situation establishes. Needless to say the possibilities were not the same for all post-colonial countries, nor were all societies automatically able to take advantage of favourable situations where they existed. A valid question to be asked in relation to the above mentioned examples of South Korea and Ghana could thus be: Were both equally exposed to the same external influences and how did these contribute to the state's capacities during the developmental process?

What is argued here is that, without taking into account the workings of the international political economy and the law of uneven development whereby internal resources are mobilized to take advantage of the opportunities offered by the external environment, the analysis of the Asian 'miracle' remains lopsided. Likewise, this analytical frame of reference ought to be taken into consideration when analysing the failure of Africa.

Post-World War II political economy

At this point it might be useful to present a brief interpretation of the workings of the international political economy since the Second World War. To a greater extent than in any other region of the Third World, this war had a revolutionizing effect on Asia. This multidimensional conflict was characterized as being three wars in one: firstly, it was an inter-imperialist conflict in the same mould as the First World War. All the industrial powers were involved in the defence or acquisition of spheres of influence and colonies. In the Pacific area, the United States and Great Britain, together with Australia and New Zealand, were involved in the repulsion of Japan's ambitions to create an economic and political zone excluding the other imperialist powers. Secondly, it was an ideological conflict between socialism and capitalism in the guise of Nazi Germany's crusade against the Soviet Union. Thirdly and most important to the present context, the conflict contained the element of being a war of national liberation against German and Italian Fascism in Europe and Japanese imperialism in Asia (Sweezy 1964: 322–4). The three levels of the conflict found their resolution in the postwar period by giving rise to three new sets of relationships:

1. The unprecedented position of the United States compared with the other industrial powers.
2. The ideological/political struggle of the Cold War between capitalism and socialism.
3. The upsurge of anti-colonialism and national liberation in the Third World generally and in the Asian sphere specifically.

Reinforcing each other, these three components of the post-1945 era became primordial in unleashing and determining the developments to come. The processes which were a function of these new sets of relationships released a chain of events that came to dominate the dynamics of international change. Developments in East Asia were in more ways than one bound up with the interaction of these three levels. Their congruence, to a higher degree than any other parameter, serves to identify the source of the transformation which has taken place in Northeast and Southeast Asia.

Following the defeat of Germany and Japan and with the weakening of England and France, American economic and political power was unchallenged. Having refused to fill the leadership vacuum during the 1930s after the failure of London as the hegemonic power, Washington was now ready to transcend the prewar situation of interimperialist rivalry. The emergence of the Cold War further accentuated the primacy of the United States. Two concepts guided the evolution of the international capitalist system under American hegemony – on the economic level: liberal international trade and on the security level: deterrence. As such, these guidelines translated into the institutionalization of a benign relationship with former foes and allies. Institutions and instruments were created to provide a benevolent trade-off whereby membership in the Western alliance would be in compliance with American global interests while bringing about benefits to the Western European countries and Japan. Thus the International Monetary Fund (IMF) and the World Bank were created as stabilizing and development institutions while the General Agreement on Tariffs and Trade would open up the way for a liberal commercial regime. The idea was that free trade would contribute to international prosperity and avoid a repetition of the economic nationalism of the 1930s. On the political–military level the creation of the North Atlantic Treaty Organization (NATO) and the entire alliance-system such as the Central Treaty Organization (CENTO) and the Southeast Asia Treaty Organization (SEATO) was meant to provide security for the allies of the United States in the face of the perceived Soviet menace and, more importantly, to offer support against domestic anti-capitalist movements and strong Communist-dominated labour unions, especially in France, Italy and Japan. China and Southeast Asia likewise were considered hotbeds of revolutionary turbulence to be neutralized.

The US strategy which was implemented in the immediate post-World War II period represented a complete reversal of the course of the postwar era which the victorious coalition had envisaged. Documents show that the intention of President Franklin D. Roosevelt towards the defeated enemies, Germany and Japan, had been to reduce the two nations to agrarian status with the objective of preventing them from unleashing future conflicts. After the demise of the American president and the emergence of the Cold War between the two wartime allies, i.e. the United States and the USSR, the new Truman administration placed great emphasis on reinstating Germany and Japan as economic centres and dynamos for European reconstruction and Asian stabilization respectively. The entire project was dictated by the concern to re-establish the capitalist international division of labour and enhance American economic interests in the world.[2]

The coincidence of geopolitics and geoeconomics

Thus in East Asia, the goal of US policy was not just the revival of the Japanese economy, but the creation of a regional political–economic construction comprising the United States, Japan and Southeast Asia. Economic interdependence was seen as providing a counterweight to the challenge of Communism in the form of Soviet influence and, more importantly, to the challenge represented by the Asian revolutionary movement. The intention behind the creation of this triangular structure was to establish a hierarchical order between the different actors: the United States as the leading unit (core), Japan as the intermediate zone in Northeast Asia (semiperiphery) with Southeast Asia as the hinterland (periphery), providing markets and raw materials for the other two units. The architecture behind this construction was conceived during deliberations of the National Security Council in Washington leading to document NSC 48 (Cumings 1984).[3] The intent towards Japan was not to transform the country into a competing centre but to keep the former enemy subordinated to American interests by maintaining US control of its economic lifeline, i.e. vital Japanese imports such as food and oil. As George Kennan, the strategist behind the plan, argued, with such an economic leash, 'we could have veto power over what she does' (Schaller 1985: 179).

Notwithstanding the efforts of the US Occupation Administration in rehabilitating the Japanese economy, it was the Korean War which put that country on its feet. American procurement orders subsidized Japan's industrial rearmament while linking its economic growth to Washington's foreign policy aims. Besides its beneficial impact on the Japanese (and German) recovery, the conflict on the Korean Peninsula ostensibly contributed to shaping the climate of the international political economy by institutionalizing the Cold War as a systemic competition between capitalism and Communism as well as an internal struggle between social forces in Western and pro-Western countries.

Not only did the war in Northeast Asia create an economic boom in the non-Communist world but it also increased the cohesion of the Western alliance. It raised the security issue to a higher level, thus strengthening these countries' dependence on American deterrence. In many respects the fate of Japan was central to US strategy. The fear of losing Tokyo to the Communist or neutralist camp after the defeat of the Kuomintang in China allowed the acceptance of internal Japanese policies which were in contradiction to American liberalism. The Korean War, besides stimulating the Japanese economy, also provided justification to the US Occupation Administration for reviving rightist political forces and big business together with reactivating the police and the military at the expense of labour and the left, who were excluded from policy-making (Cumings 1984: 20). On the macro-economic plan, the United States acquiesced in the reinstalment of a strong administrative bureaucracy (MITI) and its guidance of the economy through, among other measures, an industrial policy as well as the promotion of exports through, for instance, a weak yen. In an effort to prevent Sino-Japanese economic intercourse, the American market was eventually opened to Japanese goods. This had been a source of disagreement within the Eisenhower administration, with the president favouring allowing Japanese exports to China as a strategy for undermining Chinese socialism and protecting the US market. However, Secretary of State John Foster Dulles won the day by providing American outlets for Japan's exports.[4] With relatively easy access to US technology, the North American market, and Southeast Asian raw materials, favourable conditions for economic growth were present. At the same time, the reorganization of the Japanese labour market was conducive to keeping wages below what the gains in productivity would have warranted, thereby giving it a future additional competitive advantage on world markets (Itoh 1992).

It should, however, be noted, as Makoto Itoh points out, that until the 1960s domestic demand was an important element behind the re-industrialization of Japan, with the former colonies playing a negligible role as a market. The acquisition of postwar industrial technologies for the production of durable consumer goods induced a kind of 'Fordist regime of accumulation' with a trade-off between labour and corporate capital (Itoh 1990) which characterized industrial relations at the enterprise level rather than at the societal level. While the 'social structure of accumulation' (SSA) in the West European context was built on the welfare state, in Japan labour was promised life-long employment in the large private concerns with limited state-supported social security for the population. This translated into high savings rates.

Giving centrality to the reconstruction of Japan's economy after the 'rescue' of South Korea and Taiwan and creating the conditions for a favourable political climate in East Asia resolved a strategic problem for the United States in the 1960s while sowing the seeds of future American difficulties. Direct and covert US intervention in the area was a costly affair. The culmination of this effort was, of course, the war in Indochina, which, having demoralized and split the American people, ended in defeat. However, seen in the context of that historical period, the US military and political engagement contributed in large measure to the neutralization of revolutionary movements in the region and the stabilization of the ensuing strong right-wing regimes.

Just as the Korean War cemented the Western alliance – both on the political and economic levels – as well as deactivating the Japanese left and labour movement, the Indochina War signified the political pacification of Southeast Asia while spurring the take-off of the economies of South Korea and Taiwan. With the participation of 300,000 men over a period of seven years (1966–73), the Vietnam War played for the Republic of Korea the same role that the Korean War had played for Japan. This was a kind of 'El Dorado' for the Korean economy, accounting for approximately 20 per cent of foreign exchange earnings in the second half of the 1960s (Cole and Lyman 1971: 135). US procurement orders for the war likewise stimulated the Taiwanese economy. By the 1970s, the country was exporting capital goods and technicians and extending foreign aid to several Southeast Asian countries. In this connection it is worth noting that at the beginning of the 1960s, Taiwan had the same per capita GDP as Zaïre!

Although American public opinion is still divided on the subject of the Indochina War, it brought about a breathing space for the pro-Western capitalist political forces of this very vulnerable area. By putting 'politics in command' and carrying out military intervention, the US showed how costly revolutionary war could be for both winners and losers. Secondly, by taking advantage of the ideological conflict between Moscow and Beijing, the American engagement helped to reveal the unreliability of Soviet and Chinese support for revolutionary movements. In the 1960s, China was going through the Cultural Revolution while the USSR was promoting the policy of peaceful coexistence with the United States. This was taking place at a time when Western positions in the entire region of Southeast Asia were being challenged by social and political developments. In Indonesia, until the military coup of 1965 in which the American CIA was involved, the regime of Ahmed Sukarno, following a neutralist course, was edging toward the Asian socialist camp. Malaysia, supported by British counterinsurgency units, was grappling with a Communist guerrilla movement, while Thailand, Burma and the Philippines were facing similar insurgencies and receiving external military assistance. In Cambodia, Prince Sihanouk's neutralist stance was adapting the country's foreign policy to what seemed would be the victory of the Vietnamese Communist movement. Although it is difficult to prove cause and effect, many Southeast Asians would agree that the US war effort in Indochina bought the region a critical decade to fend off the Communist challenge and thus contribute to the vibrancy of these countries (Ott 1994).

However, the option which the Indochina conflict opened for the political elites of the region had to be capitalized on. As the then Prime Minister of Singapore, Lee Kwan Yew, expressed it in a speech in June 1966, the United States was 'buying time' for the nations of Asia and 'if we just sit down and believe people are going to buy time forever after for us, then we deserve to perish' (Rostow 1986: 14).

In other words, the global politico-military struggle against Communism which led the United States to engage itself on the Asian mainland was conducive to giving pro-capitalist elites in most countries of the region a respite in which to neutralize socialist and revolutionary movements. The strategy of the United States was thus essential to the later evolution of the East Asian 'miracle'. As an editorial of the *Far Eastern Economic Review* recently put it: 'With the benefit of hindsight, today we can see that the fall of Saigon (twenty years ago) was in fact the death rattle of a once potent communist force in Asia' (*Far Eastern Economic Review* 1994).

External interference and unintended consequences

The conjunction of foreign intervention with the internal evolution of the countries of the area is not an entirely new phenomenon. In fact, the interaction of domestic and external forces has a long history in the evolution of Asian societies. Paradoxically, most of the experiences show that outside influences often had unintended results. Seen in this light, European colonialism in Asia and the opening of Japan by the United States were essential in mobilizing social and political forces in the Japanese catching-up process. In the Southeast Asian context, Western economic and political imperialism, simultaneously and inadvertently, introduced ideological trends which later would translate into a rejection of imperialism. Thus, social humanism raised the ideal of the dignity of the individual against oppression; nationalism became a potent force against colonialism, and, closely linked to nationalism, Marxism put the struggle against imperialism and capitalism on the agenda (Dubois 1962: 42–5)

In the past century, the forced opening up of Japan together with the workings of European imperialism in Asia unleashed nationalist forces in Japan who took advantage of the international environment and mobilized internally in order to avoid the fate of colonization. Through the acquisition of Western technology and learning from the West, the Meiji Restoration (1867) implemented a 'late comer' strategy of forced industrialization and development of a strong army. Being poorly endowed with natural resources, Japan, in its efforts to 'catch up' with the West, evolved itself into a junior imperialist power and colonized Korea and Taiwan with the consent of Britain and the United States. This 'defensive' imperialism became very ambitious and, in the 1930s, under the ideological guise of Pan-Asianism (*Ajia ichi*), attempted to create a *Greater East Asia Co-Prosperity Sphere* stretching from Northern Manchuria to Southern Indonesia. Although the undertaking excluded the European colonial powers from the area, it was not an anti-imperialist project. It could best be described as a kind of contra-imperialism. Not duped by the Pan-Asian discourse, political forces in the areas occupied by Japan became mobilized in the cause of anticolonialism.

Thus, regardless of its underlying intentions, the Japanese interregnum unleashed the process of national liberation (Hewison and Rodan 1994: 245). With the defeat of Japan, the nationalist and socialist forces of Southeast Asia were not ready to accept the return of European colonialism. The metropoles who tried recolonization were defeated: the French in Vietnam and the Dutch in Indonesia. The United States, favouring the creation of a liberal world economy, discreetly encouraged nationalist forces while sabotaging socialist movements.

In this context, the victory of the Chinese Communists was essential in focusing American post-World War II political and military strategy in Asia. Besides influencing Washington's course towards Japan, this major event was the implicit reason for the US involvement in fighting two wars on the mainland which also stimulated economic growth in Japan, South Korea and Taiwan. But as an overlooked unintentional outcome, the triumph of Mao Zedong's armies over the Kuomintang itself contributed to setting in motion the wheels of capitalist growth in the region of Southeast Asia. As noted by an Indian scholar: 'Paradoxically enough, the victory of the communists in Mainland China in 1949 made a major contribution to the spread of industrial capitalism in East Asia outside Japan' (Bagchi 1987: 123). The flight of Chinese capital and businessmen to Hong Kong and Singapore resulted in the search for new forms of capital accumulation in contrast to former parasitic activities such as trade in agricultural commodities and speculation in agricultural land or usurious loans to peasants. The US blockade against China turned Hong Kong into a financial and economic centre as well as into an outlet to the world for the Chinese economy and resulted in the creation of a symbiotic relationship between the two. In addition the social experiments in China (as well as North Korea) influenced developments in South Korea and Taiwan, whereby the Americans pushed for agrarian reforms against landlordism by giving land to the peasantry. It is interesting to note that while the US encouraged land reforms in Japan, South Korea and Taiwan in order to mobilize support for modernizing political elites, there was no similar policy in Southeast Asia. The imperative of gaining support from the military and land-based overclass in that region prevented US political interference in allied regimes submitted to rural social pressures. The fear of destabilization of pro-American forces was a factor of importance.

What East Asia experienced after the end of the Second World War was a series of interrelated events which influenced the possibilities and constraints on societal development. The victory of the Chinese Communist Party over the Kuomintang ally of the United States spread Chinese capitalist elements in Southeast Asia. More important, however, was that as well as experiencing failure – first the stalemate in Korea and then the débâcle in Indochina – the US used its power and influence to re-establish Japan's economy and give it a central role in Asia. The American political and military offensive also contributed – through overt and covert support – to the establishment of authoritarian regimes in the other nations of the area. The demonstration in Indochina that revolutionary warfare was a costly affair, together with China's 'Great U-turn' and the Sino–Vietnamese conflict over Kampuchea contributed to the demoralization of revolutionary forces in the area. The post-Mao leadership in Beijing, looking at the success of authoritarian capitalism as exemplified by the 'Four Tigers', abandoned its self-centred socio-economic model. Under Deng Xiaoping, the Chinese Communist Party acknowledged the ideological and political reliability of Singapore-style authoritarianism as a worthwhile formula for its own reform process.

Authoritarian capitalism made its appearance in postcolonial Southeast Asia on the wave of anticommunism and developmental imperatives. Anticommunism suited American strategic goals while economic growth was a necessity to strengthen the legitimacy of pro-US regimes. Catering at the time to the ideological and political needs of the US establishment in its strategy towards the Third World in general and Southeast Asia in particular, American academe launched the concept of nation-building and modernity. In order to fullfil the requirements of this process, political modernization theory made the case for stability during the transformation process. Under these circumstances even authoritarian quasi-Leninist or crypto-fascist regimes were acceptable as long as they provided alternatives to compete with Communism in the Third World (Huntington 1968).

While liberal Western ideology uncomfortably accepted antisocialist dictatorships in the Third World, the Japanese offered the example of successful socio-economic and political organization in capitalist development. Asian authoritarianism distinguished itself from non-democratic regimes in other regions of the world by adopting the Capitalist

Developmental State ideology whereby industrialization could take place outside the liberal paradigm of economic and political pluralism.

The impact of Japan's political economy as a role-model – first in Northeast Asia – was further strengthened in Southeast Asia, when Japanese economic interests returned in force to the former colonial empire. On the one hand, Japanese business circles were eager to expand their foreign markets after becoming apprehensive of the potential impact of the regional integration of Europe. On the other hand, in order to fill the 'vacuum' resulting from a possible American withdrawal from Asia after Vietnam, Japan was invited to increase its trade, official aid and investment in Asian developing countries (Itoh 1990: 216–7). Furthermore the so-called Plaza Agreement in 1985, which saw the value of the yen double in the course of a few years, led to the restructuring of the Japanese economy and the relocation of manufacturing branches. These investments served as the motor for export-oriented industries in the area's low labour cost countries. In the words of a Swiss commentator: '... today the boom region forms a kind of crescent stretching from Japan deep into Southeast Asia' (Sidler 1994: 18). In addition, Japan offered these authoritarian regimes the example of successful intervention by a strong state in the economy, implicitly influencing policy-making decisions in Southeast Asian countries on the basis of slogans like 'Learn from Japan' or 'Look to Japan' (Morrison 1988: 416). Thus, having fallen under the dual hegemony of the United States and Japan, the region gained a greater political and economic leeway than other parts of the Third World. In relation to the world market the new authoritarianism which emerged in the East Asian region distinguished itself from the old authoritarianism of the Latin American-type by refusing to rely exclusively on the export of primary products and by not creating a self-reliant domestic industrialization. The regimes of the East Asian region (with the exception of the Philippines) evolved by providing stability in the transition to export-led economies with a measure of control over the various actors involved and relying on transnationals and technocrats as administrative linchpins (Woo-Cumings 1994: 414).

This evolution corresponded to the changes taking place in world capitalism. The transnationalization process leading to greater off-shore production and the search for cheap labour gave the regimes of Southeast Asia an opportunity they made good use of, thereby boosting their legitimacy and weakening oppositional forces. As noted by Hewison and Rodan: '... changes in the global political economy have facilitated a positive capitalist alternative for developing countries which has greatly undercut socialism's potential appeal in the region. One of these was of course, the search by international capital for the low cost manufacturing export bases which began in the 1960s' (Hewison and Rodan 1994: 252).

Dependent growth and middle-classization

To a greater extent than other areas of the world, Southeast Asia has been submitted to international as well as regional influences which shaped national developments. Having made the area a 'safety zone' for capital-ism, the US involvement started a process of capitalist development. Following the relative political and military withdrawal of the United States, a second wave of foreign intervention took place in addition to the American and European business presence, as Japanese corporate capital returned to the region looking for profitable investment outlets and mar-kets. Likewise, the 'Four Tigers' made their entrance in Southeast Asia in the 1980s. In addition a regional element whose role has not been entirely elucidated is the capital flows originating from the activities of the Nanyang business families. The ethnic Chinese network is an important factor in most of Southeast Asia.

The impact of external capital on internal capitalist development affec-ted the accompanying evolution of the domestic political processes. Access to foreign capital contributed to capital accumulation while sim-ultaneously relieving local business of political and administrative con-trol, in this way decreasing the autonomy of the state. Being part of the East Asian economic upsurge, it would have been extremely difficult for any regime to refuse to adapt to the demands of local, regional and inter-national capital, especially since the legitimacy of these authoritarian regimes depended on showing economic results. The 'strong state' thesis might be less appropriate as regards the leeway of the economic bour-geoisie than is usually assumed for the Southeast Asian model. It cannot, however, be denied that the state was active in the control of the socio-political evolution of civil society. As Ruth McVey warns, we should avoid looking for one-sided causality when considering the elements be-hind the capitalist evolution of the countries of the region: 'Rather than at-tempting to measure the muscles of a reified state, we can more profitably seek the source of Southeast Asian capitalist energy in the combination of factors that have encouraged business and political-bureaucratic groups to work together' (McVey 1992: 30).

To a large extent, the area can be considered to fit the pattern of associ-ated dependent development where a tripartite alliance between local and foreign capital and the state is the agent of economic growth (Evans 1988). With the exception of the Philippines the success of most countries of Southeast Asia contributed to the strengthening of non-democratic rule seen in the region. The primary factor in legitimating most of these regimes was the outstanding economic performance and a relatively stable political climate. Another element is perhaps to be found on the ideological level with the claim that their political systems are based on a different kind of democracy and value systems from the Western models. This aggressive stance has been taken by Singapore and Malaysia in particular, but other countries too, are toying with this attitude (includ-ing China and to a lesser extent Japan). A third method of legitimacy-building has relied on the ethnicity dimension, the argument being that the fragility of multi-ethnic societies requires limitations on civil liberties and controlled elections. Last but not least, by following open capitalist policies while conspicuously maintaining a radical critique of 'Western arrogance' on the issue of human rights, the regimes con-tinued to receive regional political support and international investment without weakening their nationalist credentials at home (Thompson 1993).

It can consequently be asked whether East and Southeast Asia are prototypes of autonomous development. An affirmative answer to this question would need to neglect the interaction of geopolitics and regional economics. External influences on the dynamics of the area have had a long history. Nevertheless, the understanding of the international environment's impact on Southeast Asian capitalist growth may need more research than has been done until now. In this connection it should be noted that from the beginning, the national liberation movement and anti-imperialism to a great degree determined the scope of the intervention of the United States. When Mao Zedong claimed that 'The Eastern wind would prevail over the Western wind', the menace was taken seriously by the United States. No one at the time could have predicted the outcome of the socialist challenge. But has the Western wind been so strong that capitalism is ensured a radiant future in Asia thereby preserving US and European positions there?

Another related question is whether Southeast Asia is engaged on an irreversible development path with little connection to the vagaries of global capitalism. Internally, these countries are facing difficulties created by a certain imbalance between countryside and cities, as well as by infrastructural weaknesses which will have to be alleviated in order to continue attracting export-oriented foreign capital. Furthermore, most economies of the region show a high degree of dependency on exports to the outside world where protectionist tendencies have appeared. This creates a situation which these countries cannot escape. As Ruth McVey concludes: 'The emergence of domestic business groups, their particular composition, fields of endeavour, and relationship with political power have been shaped by involvement in a world market system, from which, short of profound revolution, the Southeast Asian economies cannot secede' (McVey 1992: 33).

Besides the growth of a capitalist bourgeoisie under the new authoritarianism, most of the countries have also witnessed the middle-classization of a section of the population. The middle classes of Southeast Asia have become conspicuous and are viewed by Western mainstream liberalism as the agency of democratization and liberalization of these socio-economic formations. However, it should be recognized that a bond has been established between this segment of the population and the political-economic system. On the one hand the middle class has been moulded by this special form of state capitalism by being exposed to a national education system, and on the other hand by depending on employment in the various branches of the techno-bureaucracy and the local capitalist enterprises. The dynamism and interests of both bourgeoisie and middle class are linked to political stability seen as central to export-oriented growth. Ironing over Western expectations of the withering of a redundant authoritarianism on the basis of middle-class politics, an analyst based in Singapore puts the question in the following way: 'In other words, the middle class, itself the frankensteinian creation of the Asian developmental state, constitutes the liberal democratic nemesis. But does it?' (Jones 1994/95: 46). The gist of his article is that it does not.

Conclusion

On the basis of this analysis, the thesis is advanced that the coincidence of geopolitics and geoeconomics was favourable to the internal socio-economic capitalist processes of Southeast Asia. But this evolution was not inevitable. For example, the case could be made that without the victory of Mao Zedong, the focus of US interest in the region might well have been different, perhaps concentrating on China at the expense of the rest of East Asia – including Japan.

What is being argued here is that in order to understand the evolution of East Asian societies it is necessary to integrate different levels of analysis and not least to grasp the international political-economic dimension. Without the interference of the United States in the affairs of the region generally as well as in single countries specifically, the socio-economic and political processes would, in all probability, have taken a different course. Furthermore, the favourable conditions existing on the world market made the export-oriented industrialization strategy of most of these authoritarian regimes possible. Thus access to Japanese mid-range technology and capital as well as the US market played a primordial role.

This ought not to be interpreted as meaning that other levels of analysis should be discounted. But in the world system there is a great degree of interdependency which gives possibilities to some while constraining others, depending on the interests of the core nations and especially those of the leading power.

Where does all this leave us? Can projections be made into the future? History has not ended, especially in Southeast Asia where the next challenge will be the integration of the Chinese economy in the region. Whereas, until the 1980s the People's Republic of China was seen in Southeast Asia as an anticapitalist menace, making economic growth as a necessary counter-weight, China today is becoming more of a competitor to its neighbours for Western technology, capital and access to external markets, especially those of the United States, Japan and Europe.

The contest will now be one of economic competition and of adjustment of internal structures to external developments within the context of capitalism. This implies the need to give renewed emphasis to the internal level of analysis, i.e. socio-economic and political aspects, especially now that, in addition to a large working class, a substantial middle class has appeared on the heels of the capitalist evolution. This heterogeneous class is perhaps, as seen above, not as anti-establishment (anti-authoritarian) as mainstream liberalism would like it to be. Nevertheless, it has probably developed a split personality with, as some see it, its own agenda and demands for political freedom and economic wellbeing (Hewison and Rodan 1994: 252–6). However, it is important to determine whether the priority of the middle class is economic wellbeing with law and order or political democracy with political instability accentuated by economic insecurity. Until now the 'virus' of democratization, which has affected Latin America, Africa, Eastern Europe and Russia, seems to have by-passed the 'nouveau-riche' states of the Asian sphere. As Mark A. Thompson notes: 'If ASEAN's non democracies have prospered politically even during the hightide of democracy, what will a "reverse wave" of democratization in the so-called Third World bring? The Asian model of development dictatorship may become a kind of alternative political model to Western democracy' (Thompson 1993: 482).

Furthermore, experience shows that domestic problems can lead to external frictions, especially where nationalism is still a potent force. In this connection it should not be forgotten that Southeast Asia still has unresolved territorial disputes. In the longer run, the internal and external processes at work will again translate into the issue of security. It is no coincidence that the countries of Southeast Asia appear to be involved in a rather intensive arms race. In addition, the later frictions in the US–Japanese partnership do not alleviate the genuine anxieties concerning the stability of the region.

The world has been fascinated by the economic growth of East Asia in the past two decades. The region's dynamism impresses and raises fears in the West. Northeast and Southeast Asia will most probably continue to capture the attention of the rest of the world in the next century, but much will depend on the interaction of the international, regional and internal environments. Will the domestic forces in each country be able to influence the other two levels, or will the individual societies remain at the mercy of external processes?

Before concluding, a word of caution. The American economist Paul Krugman reminds us in a recent article in *Foreign Affairs* that not long ago the world was alarmed by Soviet industrialization and rates of growth. Naïve projections of growth rates of the Warsaw Pact nations in the 1950s into the future overstated the real prospects. Is the same error not being repeated with regard to the Asian boom? His argument is based on economic studies of East Asia showing that the high-growth era was driven by unusually high inputs such as labour and capital rather than gains in efficiency (Krugman 1994). In any case, it can be expected that the need to 'squeeze the lemon' in order for these economies to remain export-oriented will bring different forms of social contradictions to the fore again. In this relation it is highly doubtful that the middle classes can remain sheltered.

It is ironical in the context of the history of East Asia, to hear Western admonitions against Asian-style non-democratic regimes. This discourse in its most extreme form has been launched by the foremost American proponent of authoritarianism, Samuel P. Huntington, who thirty years later sees in the dichotomy of Western type-societies/non-Western and non-liberal type-societies a 'clash of civilizations' (Huntington 1993).

1. The following passages were inspired by Gourevitch (1989: 11–12).

2. See chapter 1 in Hersh (1993).

3. Also reproduced in Deyo (1987).

4. For a brief discussion of this episode see Hersh (1993: 29–32).

Bibliography

Bagchi, Amiya Kumar (1987) 'East Asian Capitalism: An Introduction', *Political Economy*, **3**(2)

Cole, David C. and **Princeton N. Lyman** (1971) *Korean Development: The Interplay of Politics and Economics*, Harvard University Press, Cambridge, Mass.

Cumings, Bruce (1984) 'The Origins and Development of the Northeast Political Economy: Industrial Sectors, Product Cycles and Political Consequences', *International Organization*, **38**(1) Winter

Deyo, Frederic C. (ed) (1987) *The Political Economy of the New Asian Industrialism*, Cornell University Press, Ithaca, NY, and London (second edition 1988)

Dubois, Corra (1962) *Social Forces in Southeast Asia*, Harvard University Press, Cambridge, Mass.

Evans, Peter (1988) 'Class, State and Dependence in East Asia: Lessons for Latin Americanists', in Frederic C. Deyo (ed.) *The Political Economy of the New Asian Industrialism*, Cornell University Press, Ithaca, NY, and London, second edition

Far Eastern Economic Review (1994) '20 Years of Asian Growth', *Far Eastern Economic Review*, 24 November

Gourevitch, Peter A. (1989) 'The Pacific Rim: Current Debates', *The Annales*, September

Hersh, Jacques (1993) *The USA and the Rise of East Asia*, Macmillan, Basingstoke and London, and St Martin's Press, New York

Hewison, Kewin and **Garry Rodan** (1994) 'The Decline of the Left in Southeast Asia', *Socialist Register 1994*, Merlin Press, London

Huntington, Samuel P. (1968) *Political Order in Changing Societies*, Yale University Press, New Haven, Conn.

Huntington, Samuel P. (1993) 'The Clash of Civilizations', *Foreign Affairs*, July/August

Itoh, Makoto (1990) *The World Economic Crisis and Capitalism*, Macmillan, Basingstoke and London

Itoh, Makoto (1992) 'Japan in a New World Order', in Ralph Miliband and Leo Panitch (eds) *Socialist Register 1992*, Merlin Press, London

Jones, David Martin (1994/95) 'Asia's Rising Middle Class – Not a Force for Change', *The National Interest*, Winter

Krugman, Paul (1994) 'The Myth of Asia's Miracle', *Foreign Affairs*, November/December

Marx, Karl (1958) 'The Eighteenth Brumaire of Louis Bonaparte', in Karl

Marx and Friedrich Engels *Selected Works*, **1**, Foreign Languages Publishing House, Moscow

McVey, Ruth (1992) 'The Materialization of the Southeast Asian Entrepreneur', in McVey, Ruth (ed.) *Southeast Asian Capitalists*, Southeast Asia Programme, Cornell University, Ithaca, New York

McVey, Ruth (1995) 'Change and Continuity in Southeast Asian Studies', *Journal of Southeast Asian Studies*, **26**(1) March

Morrison, Charles E. (1988) 'Japan and the ASEAN Countries: The Evolution of Japan's Regional Role', in Takashi Inoguchi and Daniel I. Okimoto (eds) *The Political Economy of Japan*, **2**, *The Changing International Context*, Stanford University Press, Stanford, Calif.

Ott, Marvin (1994) 'For Southeast Asia, a Crucial Respite', *International Herald Tribune*, 29 April

Rostow, W. W. (1986) *The United States and the Regional Organization of Asia and the Pacific: 1965–1985*, University of Texas Press, Austin

Schaller, Michael (1985) *The American Occupation of Japan – The Origins of the Cold War in Asia*, Oxford University Press, Oxford and New York

Sidler, Peter (1994) 'Swift Economic Growth in Southeast Asia', *Swiss Review of World Affairs*, November

Sweezy, Paul M. (1964) *The Theory of Capitalist Development*, Monthly Review Press, New York (5th edition)

Thompson, Mark R. (1993) 'The Limits of Democratization in ASEAN', *Third World Quarterly*, **14**(3)

Weber, Max (1968) *Religion of China*, The Free Press, New York

Woo-Cumings, Meredith (1994) 'The "New Authoritarianism" in East Asia', *Current History*, December

World Bank (1991) *World Tables*, World Bank, Washington D.C.

CHAPTER 2

The custodian state and social change – creating growth without welfare

Johannes Dragsbaek Schmidt[1]

The relationship between politics and the societal realm in the process of capitalist modernization is both complex and intriguing. In recent years increasing attention has been given to the implications of state restructuring and state–society relations in general, and to patterns of economic growth as well as the consequences of these transformations for income distribution and welfare. A number of scholars have focused on administrative structures promoting growth, generating useful analytical constructs, such as that of the developmental state.[2] Other works emphasize the role of private sector organizations and interpersonal networks supporting the developmental state or how the workings of market institutions are facilitated by this particular political entity.[3] However, the broader political correlations between internal and external politics of state structures remain relatively unexplored.[4]

The aim of this chapter is to examine the interaction of social change with distributional problems stemming from state policy-making and market allocations in one of the fastest growing regions of the world. It attempts to discuss the relationship between institutional capacities of state structures in the ASEAN-4 (Indonesia, Malaysia, Thailand and the Philippines) in relation to strong growth and the uneven spread of benefits. The introductory section is devoted to a theoretical presentation of emerging welfare functions of government and the role of the market. The second and third sections give a historical analysis of the institutional policy set-up and a description of the relationship between growth and equity. The fourth section focuses on economic policies and their concomitant uneven consequences. This aspect of non-welfare is further elaborated and the argument is finally outlined that a comparative perspective on internal and external constraints and possibilities of improving welfare and income distribution in the ASEAN-4 shows that contradictions between political will (government) and actual implementation (capacity of the bureaucracy) together with societal pressures shape the outcome of welfare.

Political will and the market

The political economy of capitalism, as its keenest critic Karl Marx observed, generates a rapid long-run rate of capital accumulation and a tremendous increase in the forces of production. Marx also stressed, however, that the development of capitalist economies tends to be very uneven, both spatially and temporally; there are good and bad times. Indeed, most capitalist societies have experienced alternating periods of rapid economic expansion lasting several decades, as well as periods of generalized economic crises lasting a decade or more. Each such long-run boom and subsequent crises constitutes a long cycle. Furthermore, each long cycle in the development of a capitalist economy can be associated with a historically specific social structure of accumulation.

How do these observations relate to recent developments in Southeast Asia? The changes in the region's economies over the last thirty years have been spectacular. Substantial industrial sectors have been established and, in all countries except Indonesia, the former dependence on primary exports has been reduced. These rapid transformations were made possible by the evolution of the international economy and have more or less followed the historical up- and down-turns of modern global capitalism: the crisis in the 1930s was followed by substantial economic growth, and again the crises in the 1970s and first half of the 1980s were followed by a period of restructuring and then rapid growth. What is quite significant for these two periods of growth is the importance of factors external to the economies of Southeast Asia.[5] This fact ultimately confronts a concrete problem, namely one of strengthening the productive forces and the combined high economic growth in Southeast Asia in a manner which is socially and politically beneficial to the region's populations, especially to the poor and marginalized segments. If capitalist development is deepening in Southeast Asia and markets in their own right are emerging, it remains questionable whether this might be considered as an autonomous capital development alternative.[6]

Non-ideologically motivated social science agrees on the important point that the market system – capitalism – is not a self-regulating entity. It does not by itself add up to a socially defensible allocation of either private income or public investment. It does not efficiently or fairly distribute certain necessary social goods such as education, health, roads or research spending. Left to their own devices, markets do not broker social contracts, which are needed, even in narrow economic terms, to compensate for the shortcomings of the economic sphere. In fact, social contracts and markets coexist within a conflictual rationality where the state becomes the crucial battleground. Indeed, the argument that follows stresses that social groups challenge the role of the state as a provider of collective and distributional goods, and as a repository, creator and mediator.[7] At the same time, the particular relationship with the international political economy influences policy options and functions as a direct and indirect pressure on the state's ability to respond to societal pressures.

Ascribing such central significance to the conflictuality of social objectives means that, ever since political activities became concentrated in an entity separate from the rest of society, the existence of the state has implied a claim on resources. In the realm of market intervention the state appears as an economic player and as a regulating agency. The derivative role of the state as an economic agent in the domestic context acts with a triple objective: as a producer of public services and goods, as a regulating agency, and as manager through direct intervention in private sector economic activities. State services include the provision of public goods while state participation in economic development comes in two forms: direct participation via public ownership of enterprises and indirect participation via fiscal and trade policy.

In Southeast Asia, the role of the state fulfils both the function of *accumulation* and the function of *legitimation*:[8] the first requires aid to capital and control of labour, the other accelerates economic growth. The accumulation function is related to the government's supervision of the rate and structure of gross investment which includes forcing the pace of savings, plugging any shortfall in the rate of domestic investment funds (private or public) and changing allocations between industries so as to achieve predetermined goals. It also includes intervention in the labour market in order to underpin the investment strategy. The legitimation function implies keeping discontent in check by providing reasonable improvement in living standards, pacifying capital–labour conflicts and attempting to maintain social cohesion. The legitimation function is closely intertwined with the accumulation function since the latter fundamentally affects changes in welfare and consumption patterns.

Where the state becomes both the engine of development and the arbiter of social relations, benefits are distributed by 'concessionary' bargaining between local and foreign actors and institutions. Over time, state-controlled dispensation of 'favours' becomes important for the maintainance of the regime. In order to understand the specific Southeast Asian model of welfare and its position between the role of the state and social change, a discussion of methodology and conceptual design is needed.

Policy-making, institutions and social change

Historical and theoretical studies show that economic growth itself tends to exacerbate income inequalities. There is no automatic trickling down of welfare accompanying it, but rather a trickling up. Any improvement in the distribution of income requires a deliberate course (Martin 1991: 47–8).

The aims of economic policy are sometimes formulated as a social welfare function to be maximized (or minimized), subject to certain constraints. The social welfare function comprises target variables, and, possibly, some instrument variables. Physical or political constraints can limit the permissible range of instrument variables – a state of affairs which must be taken into consideration in the designing of policies for fixed targets.

In functionalist terms, a successful economic policy is one that secures a welfare system which sustains and develops the labour force. This has a number of advantages. It buys social peace by blunting the impact of distributive conflicts which are at the heart of capitalist development. This manifests itself in lower levels of strike activity which is considered to be a major source of loss of economic welfare and capitalist efficiency. More positively an educated and healthy population is a more efficient work force; no country can promote technological change and industrial innovation with a dilapidated social infrastructure, an inadequate educational system, poor housing and high levels of economic insecurity.

There is a whole range of alternative policy instruments for alleviating income inequality. Policies must ensure that the purchasing power of the low-income group is increased, that the supply of goods is sufficient for basic minimum consumption standards, that the available supplies actually reach the target groups, and that the redistribution policy is sustainable. Policy instruments of redistribution come in various forms: price control, wage guidelines, taxes, employment creation, public goods and redistribution of assets. However, the intention here is not to undertake a complete review of all these complex policy issues, but to develop the argument by focusing on the specific interaction of states and societal change.

The centrality of the state as an institution and actor on its own premisses in the politics and economics of development is manifest. The following general features apply to the institutional set-up in the ASEAN-4: the degree of autonomy and insulation of technocrats and other policy-makers have determined the relationship between state priority on growth and the neglect of social welfare distribution, (except in Malaysia). This is partly due to the degree of stable bureaucracies and in all countries concerned, with the exception of Thailand, long-term stable regimes (Marcos, Suharto and Mahathir) which created the necessary political stability, and finally the exclusion of the majority of the populations from political participation and influence on important policy questions.

With these general points in mind, it is useful to examine the complementary question of growth and welfare at a more detailed level.[9] This will be done by comparing the political aspects of growth and welfare in Indonesia, Malaysia, Thailand and the Philippines.

Growth and equity in Southeast Asia

The dominant political tensions within capitalism concern the control of investment resources and the division of the economic surplus. This is also the case in Southeast Asia where historical confrontations have occasionally revealed the mobilization of excluded groups and classes.[10] Ostensibly, the state's interest in maintaining harmonious labour relations has been related to its tasks of fostering an attractive investment climate for purposes of national accumulation in addition to important security concerns referring to the legitimation function. To maintain harmony, trade unionists are officially enjoined to accept the need to restrict their activities in the name of the 'national interest'. One example is the dispute of the Malaysian Ministry of Labour with the ILO (International Labour Organization) concerning the creation of a peaceful climate of industrial relations deemed to be essential for growth (Alagandram and Rahim 1988: 11). Similar situations have occurred in Thailand, Indonesia and the Philippines.[11] This aspect of political stability provides one important explanation for the exceptional combination of high economic growth and weak distribution of social and political goods.

As can be discerned from Table 2.1, the ASEAN-4 represent the medium performers in the Asia-Pacific region, next to the NICs and Japan, but ahead of South Asia. The growth of GNP per capita of this group averaged 3.5 per cent during the 1965–90 period, with the Philippines showing the poorest performance (1.3 per cent) and Indonesia and Malaysia, buoyed by their oil wealth, more than doubling the growth rate of the Philippines. Thailand's GNP growth rate was also more than double that of the Philippines. For Thailand and Malaysia, growth provided an exceptional flow of resources for development and social welfare. Even the Philippines enjoyed substantial growth until the domestic political crises and world economic recession in the beginning of the 1980s.

Indeed several measures of social welfare aggregate do show substantial improvement. Through the three decades from the 1960s onwards, Thailand had some 9 million absolute poor and the Philippines 13 million. However, these countries succeeded in reducing the incidence of absolute poverty by about a third, to 16 per cent in Thailand and 21 per cent in the Philippines. Nevertheless, in 1990, the incidence of poverty in Thailand was as high as in Indonesia even though its average GNP was 2.5 times higher. In absolute numbers about 50 million remained poor. If we look at the UNDP's (United Nations Development Program) Human Development Index (1993), there was a general increase of general wealth parameters at almost all levels.

Using the conventional yardstick of aggregate or per capita income growth, the economic performance of the resource-rich ASEAN-4 in the

Table 2.1 GNP per capita growth and absolute poverty in the ASEAN-4, 1970–90

Country	Average annual growth rate (%) 1965–90	Poverty					
		Incidence (%)			Number (millions)		
		1970	1980	1990	1970	1980	1990
Indonesia	4.5	60	29	15	70	42	27
Malaysia	4.0	18	9	2	2	1	0.4
Philippines	1.3	35	30	21	13	14	13
Thailand	4.4	26	17	16	9.5	8	9

Note: Due to differences in coverage and definitions, cross-country comparisons have their limitations.
Source: Sustaining Rapid Development in East Asia and the Pacific. Development in Practice Series. The World Bank 1993.

last decades can be said to have been impressive. However, growth does not correspond to the welfare level when broken down to measurements such as income distribution, and social and regional differences. This situation is revealed by the political correlates of economic growth such as the type of regime and the choice of development strategy which concomitantly have important distributional consequences for the society. The ASEAN-4 adopted a non-democratic form of regime aimed at maximizing national accumulation through industrialization strategies based on import substitution and export orientation. Though the timing was different in each country, these choices put the rural sector in second place on the list of priorities.

It is recognized in development studies that capitalist economic growth goes hand in hand with a relative level of exploitation and inequality. This increasingly uneven development does exact a high social cost from the population, which ultimately can lead to economic disarticulation, social tension and political unrest. In most instances, the unevenness of growth in the ASEAN-4 economies has aggravated social inequality.[12] Regional inequalities have also increased and disparities among and between classes and social strata in the rural as well as in the urban areas are growing.

Not only in the countryside, but also in the metropoles increasing disparity in living conditions and environmental degradation are shown. In addition there is a growing gap between the relative improvement in income generation and the acute needs in collective consumption and urban services. Housing, health, public transport and social services do not keep pace with the growth of social and functional needs of the city population. As Castells points out, 'the urban crisis becomes an important aspect of the social blight which is an integral part of the industrialization process in the ASEAN region' (Castells 1991: iii–vi). Furthermore, a widening gap exists with respect to per capita incomes in the metropolitan centres relative to those of other regions.

The problems are also glaringly apparent in terms of the differential incomes of the wealthy and the poor. On top of the income pyramid is a small upper class consisting of a disproportionately high percentage of pure Chinese or mixed Chinese-indigenous ethnic strain in the four countries where this small minority comprises a fraction of the population. Included among the very rich of non-Chinese origin in Thailand and Indonesia is a small number of top military officers and bureaucrats, either in active service or retired. They were invited to serve on boards of directors of several locally- and/or foreign-owned industries and businesses, and were either given or sold stocks in these enterprises at artificially low prices: some of these officers, after amassing sufficient gains over the years or after receiving preferential credit from government banks, eventually started their own enterprises (Thomas 1991: 14). This segment dominates the society in nearly all respects. With only few exceptions these groups set societal norms and control most of the wealth and the political system at the expense of the marginalized strata.[13] As shown below, only Malaysia followed a more or less successful policy of eliminating poverty.

In Thailand, the richest 20 per cent households earned about 49.3 per cent of total income in 1975/76, with the proportion rising to about 54.9 per cent in 1987/88. The poorest 20 per cent earned about 6.1 per cent in 1975/76, falling to 4.5 per cent in 1987/88 (Ratanakomut et al. 1994: 205). The situation in Indonesia reflects no change. The richest 20 per cent earned 42 per cent of total income in 1980 and likewise about 42 per cent in 1989. The poorest 40 per cent earned 20 per cent and saw a slight increase to about 21 per cent in 1989 (Djojohadikusumo 1989: 51). In the Philippines, the richest 20 per cent earned 54 per cent of total income in 1970/74, but this estimate decreased slightly to about 52 per cent in 1986/89. The poorest 30 per cent earned 7.1 per cent in 1970/74, but only increased to 9.3 per cent in 1986/89 (Lamberte 1992: 346). In Malaysia the share of the top 20 per cent declined from about 55 per cent in 1973 to about 52 per cent in 1987. The share of the bottom 40 per cent increased from 12 to about 14 per cent in the same period (Bhalla and Kharas 1992: 47).

Deterioration in the social distribution of income and wealth represents an important potential fetter on economic growth, which can also play a role in sparking the downswing of the business cycle. The growth of real GDP has delivered social benefits mainly in the form of new employment opportunities rather than in a general improvement of living standards for the bottom 20 per cent of the income pyramid (McFarlane 1988: 33).

The uneven consequences of a non-democratic regime and elite policy-making

In Southeast Asia, there has been a marked tendency to avoid serious discussions of the social disruptions of the growth process; academic writings and policy proposals usually treat actual or potential conflicts as a matter of political technique. During the 1960s and 1970s, left-wing forces mobilized labour in rural and urban areas against what was termed the evil of underdeveloped capitalism.[14] The socialist movement in the region was, as documented by Jacques Hersh in this volume, crushed by US direct and indirect interventions, fatally weakened as a consequence of the Sino–Soviet split and, more importantly, the Sino–Vietnamese conflict. Together with the beginning of economic growth under strong governments, socialist movements were defeated.[15] This translated into giving the state elites a greater leeway than would have otherwise been the case.

Although the situation in the Philippines is treated as more similar to the Latin American model of strong interference in economic policy-making by well-connected business groups, the general pattern in the ASEAN-4 is one of dominance by a more insulated technocracy. The role of non-elected, nominally apolitical policy-makers, i.e. technocrats, responsible for formulating and carrying out national economic policies, is an aspect of policy formulation common to all four countries. Another common feature is the close relationship between technocrats, economists, and select academic and research institutions. Technocrats in Indonesia, Malaysia and Thailand have remained relatively insulated from demands for accountability and participation – although in Thailand, some banking and business interests have proved able to break the autonomy of economic policy-makers. By comparison, in the Philippines technocrats lost power, temporarily, during the transition from the Marcos to the Aquino regime (Schmidt 1993a: 29; Schmidt 1993b).

Considerable contextual evidence suggests that Southeast Asian state officials, because of personal socialization and material benefits derived from connections with TNCs (transnational corporations) and related elites from the US, Japan and Western Europe, have substantial motivations to repress disruptive local labour in TNC-enterprises. Both formal and informal socialization mechanisms have been found to instil in state officials a developmental preference for TNC-based production. The Economic Development Institute of the World Bank, for example, forms technocrats through TNC-dominated export-oriented investment programmes. Similarly, educational courses are sponsored by the Asian Development Bank to induce market-conforming methods by teaching deregulation and liberalization. Given the lending leverage these institutions have over technocrats, one must assume that the instruction is taken seriously. Likewise, a great number of Japanese and American foundations provide essential funds for education and scholarships for students, researchers and bureaucrats. Indeed, university education in the North has given birth to such appellations as the 'Berkeley Mafia' in Indonesia and the 'Ivy League Mafia' in the Philippines.[16]

Liberalization-oriented stabilization policies and deregulation implemented in the ASEAN-4 in the last decades had mixed results in terms of their objectives, and were detrimental to growth and income distribution. The ambiguous achievements in terms of stated objectives (containing inflation and rectifying chronic balance-of-payments imbalances) show that the standard policy packages introduced by the IMF and the World Bank are not well designed and do not correspond to the realities of the societies concerned. One reason is that strong vested interests in each country, both among the bureaucrats and a nationalist-oriented fraction of the bourgeoisie occasionally succeed in resisting these policies.

The nation-building strategies of the ruling elites of Southeast Asia tell us that a policy combining various incentives (fiscal, financial, exchange rate and trade incentives – including protectionism) forces industrialists to export. The explanation is that a developing economy cannot overcome its technological disadvantages of late industrialization without the aid of subsidies, protection and coordination by the state (Amsden 1989: 13–14).

Industrial protection and import-substitution are policy instruments that raise the domestic price of competing import products relative to those exported. As can be seen from Figure 2.1, the effective rates of protection (EPR) in manufacturing are high and, as pointed out elsewhere they actually increased during the 1980s (Schmidt 1994). A marked difference between the Northeast Asian model and the Southeast Asian one, however, is to be found in the societal fallouts of the process in Southeast

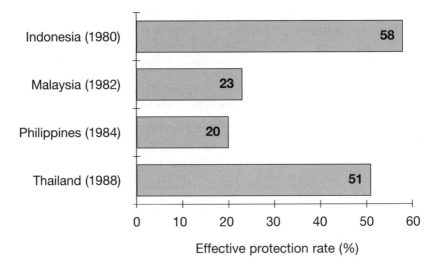

Figure 2.1   Effective protection rates in manufacturing in the ASEAN-4
Source: Sustaining Rapid Development in Asia and the Pacific.
Development in Practice Series. World Bank. 1993.

Asia. Whereas the patterns of industrial and trade policies in Japan, Taiwan and South Korea had relatively benevolent implications for the distribution of public services and welfare between rural and urban areas, the opposite was the case in the ASEAN-4.

In Southeast Asia, the uncontrolled rapid growth of the metropolitan centres has exceeded the capacity of the state bureaucracies to provide essential public utilities and services. Blackouts, floodings because of lack of piped water facilities and ground water, lack of sewerage systems and increasing pressure on rubbish collection, heavy pollution and traffic congestion have all led to chaotic conditions and growing public discontent with state and metropolitan policies.

Urban-biased development policy has been in operation to varying degrees in Indonesia, Malaysia, Thailand and the Philippines, where roughly a quarter to one-third of the urban population is found in the metropolitan capitals. The comparatively low urban concentration in Indonesia must be related to its large territorial expanse. This can hardly be credited in any significant measure to Indonesia's transmigration programme which has had only limited success. The relatively moderate urban primacy in Malaysia is the result of slower rural–urban migration, which in turn has been made possible by the rural thrust of its development policy since the late 1960s, including the promotion of regional growth centres since the mid-1970s. A major policy instrument in Malaysia has been the development of land schemes (FELDA) supplemented by other measures such as agricultural extension services, provision of credit and marketing schemes, drainage, irrigation and transport facilities.

In the region as a whole the rapid growth of economic activity and of population continues to be concentrated in the metropolitan capitals. This is the consequence of economic and social policies that have implicitly favoured large cities at the expense of rural areas and small provincial towns, implying likewise that explicit spatial distribution policies have been largely ineffective (Pernia 1991: 22). For the capitalist countries of Southeast Asia, any critical evaluation of policies adopted specifically for the purpose of influencing urbanization is likely to reach the conclusion that these policies are relatively weak instruments compared with the basic and powerful effects of certain underlying trends and of government policies adopted with other ends in view. This statement could well have been made with respect to attempts at reducing the unevenness of development as a whole.[17]

While state intervention with financial aid and advice from the World Bank favours the urban centres, it has, to a certain degree, supported the rural sector in terms of investments in irrigation, transport, electrification, health and education that help to overcome some production constraints. However, it is in the area of agricultural pricing policies that state intervention in the ASEAN-4 has not often worked to the benefit of farmers.[18] Prices of many farm products, especially export crops, have been suppressed through the operation of marketing boards, the imposition of explicit taxes, and direct price controls. While subsidies for agricultural inputs (fertilizer, irrigation water, credit) have been provided, mostly to food crop producers, they do not always compensate for the low prices of farming produce. Even worse from a distributional viewpoint, these input-subsidies have tended to favour the large landholders. Except for rice in Malaysia, which became heavily protected during 1975–83, negative values for the total nominal protection rates, combined with overvaluation in real exchange rates, have on several occasions exerted a negative effect on agricultural production incentives in Thailand and the Philippines. As Bautista notes, 'such anti-agricultural price bias of government policies, perhaps unintended in many cases, is presumptive evidence that farm production and exports would have been larger under a less restrictive foreign trade regime. It reduces the effectiveness of public investment in agriculture, nullifying at least in part the benefits derived from it by agricultural producers' (Bautista 1992: 36–7). The upshot is that the net effect of the price policies might have contributed to an increase in poverty and its negative output effect have given rise to an effective resource drain out of the rural sector.

The public distribution of (subsidized or free) foodstuffs is a close alternative to income transfers. But the implementation of these schemes may have increased the inequality of consumption. In Thailand, the distributed supplies have been available mostly in urban centres. Since city incomes are usually higher than rural incomes, the urban favouritism of the food distribution systems may have accentuated inequality.

The urban bias in development policy has been reinforced by the degree of openness to the world market. As the following shows, the level of welfare is also determined by the economic strategy each country has adopted. This is particularly so in the case of investment and labour policies.

The state as conflict arbiter and caretaker of outward-orientation

Regarding FDI (foreign direct investment) from TNCs, the voidance shows that they tend to be relatively insensitive to social concerns. Thus, TNC workers in the ASEAN countries are in general debarred from participating in independent labour union activities; they are effectively prevented from having autonomous organizations defending their standards of living through collective action. TNC workers are almost wholly dependent upon the goodwill of management. Recent data shows that all the states in the ASEAN-4 intervened directly in slightly over one-half (51.9 per cent) of labour actions between 1968 and 1983. Workers experienced firings in 49.0 per cent of the events; detentions in 12.2 per cent; injuries in 8.2 per cent; and coerced terminations in 32.7 per cent. At least one form of repression occurred in two-thirds (67.3 per cent) of the cases.[19] But they are still better off than those in the usually non-organized rural sector, as can be seen from the problems caused by internal migration to the cities.

In the last decade ASEAN governments made use of various policy instruments such as wage constraints toward unions with the aim of achieving stabilization rather than even redistribution. The suppression of trade unions for the purpose of export-oriented industrialization caused the erosion of real wages, but the organized industrial working class was also weak, with a restricted political potential and mobilization. With a combination of brute force and labour regulation policies it was relatively easy for the state elites to establish domination over and to incorporate public trade unions which were the strongest in terms of membership, numbers and strength. This is related to the tradition of the pre-independence period. During the colonialist era in Malaya, the British fostered the formation of the Malayan Trade Union Congress (MTUC) as an alternative to the Communist unions; the MTUC today continues to be a moderate organization, hemmed in by many restrictive laws. In Indonesia, a government-sponsored labour federation replaced independent unionism, which collapsed with the crushing of the Communist Party in 1965. Unions in Thailand have traditionally been small and fragmented, and occasionally under the influence of factions in the military and the government. Only in the Philippines does the leftist labour movement represent some sort of challenge, but it is still not a potent force (Crouch and Morley 1993: 285).

The lack of bargaining power of the working class is also accentuated by unemployment and widespread underemployment in Thailand, Indonesia and the Philippines. Considering the sectoral distribution of production structures and the labour force, it is significant that none of the countries have experienced a substantial shift to agro-industrialism. The proportion of the work force in manufacturing and construction is only 13 per cent in Indonesia and the Philippines, 15 per cent in Thailand and 14.5 per cent in Malaysia. Omitting Malaysia, the share of the labour force in agriculture in these countries is more than half, thus showing a much lower level of employment transition and mobility.

As shown above, the problems of economic policy are difficult to analyse in depth without taking into consideration the political and institutional environment as well as the constraints on policy-making. The lack of experience and expertise is a dilemma that plagues policy-makers throughout Southeast Asia. Even with the best of intentions, the deficiency of certain skills can be a formidable obstacle to policy-making. The instruments at the disposal of a modern state are numerous and there are trade-offs and ramifications that can confound any carelessly designed policy. The limitations in the managerial capabilities of the state are frequently used by the World Bank and IMF as an argument against interventionist policies. There is no doubt that these weaknesses must be taken into consideration in the design of policies (the selection of instruments, determination of monitoring methods, etc.). But these shortcomings need not determine the aims and priorities of the state. Even the efficient supervision of a regime of laissez-faire requires managerial skills and heavy regulation.

Regardless of the difficulties of implementation, it is necessary to focus on the political regime and the choice of development strategy to understand the processes these nations have followed. This is possible by distinguishing between 'political capacity' and 'political will' as the two primary variables that affect the non-market distribution of social welfare. 'It is apparent that both an ability to implement change (political capacity), and the willingness to do so (political will) are requisites of public-initiated welfare changes; neither alone is sufficient' (Crone 1993: 56). According to this, political variables like capacity and will, respectively, derive from the underlying structure of the political system and are rooted in intra-elite struggles over social change. Together they constitute different responses in either a welfare-enhancing or welfare-inhibiting manner. Thus, the critical elements in state capacity rest on a broader pattern of socio-economic elite support; not democracy. Furthermore, political structures facilitate or hinder redistributive reform by providing more or less capacity to act autonomously. Crises stimulate and call for change.[20]

The situation in Malaysia confirms this thesis. In an evaluation of the New Economic Policy, formulated as a response to the 1969 crises, Prime Minister Mahathir made the point that, by emphasizing basic services and redistributive control of economic resources, the programme had eradicated acute poverty, raised the general welfare, and achieved growth with equity.

Likewise in Thailand the so-called Accelerated Rural Development Program (ARDP) was essentially established as a response to the brief democracy period of 1973–76.[21] The importance of the ARDP was stressed by King Bumipol, who expressed concern that 'our own survival depends on the stability and security of the rural people'.

The rhetoric of crises and reform is also a constant of Philippine politics. As a consequence of an ineffective martial law and crony capitalism under Marcos, it became clear that the endeavour to restore legitimacy of the Filipino state was not possible. The new Aquino regime came to power with a promise to implement a comprehensive programme of agrarian reform. However, as in the case of Thailand, vested interests in the military and big business blocked its implementation (Jayasuriya 1992: 62, 84).

In Indonesia, it was the change, after a brief but extremely bloody period, from the Old Order to the New Order which marked the establishing of a comprehensive reform programme. One of the outcomes was major contradictions in the political structure of the New Order between the Bappenas technocrats and the military-dominated politico-bureaucratic factions over economic policy and more specifically control over the resources of the state (Robison 1986: 140).

Crises pushed leaderships towards reformist programmes in the ASEAN-4, except in the Philippines where vested interests located outside the state apparatus were too well entrenched. The higher state capacity of Malaysia provided it with the means, the crises with the motivation, and the intra-elite struggles with the mode of reformist change. While the Filipino state experienced crises, it has had neither the capacity nor the strategic leadership to carry out reforms. In Thailand, evidence shows that the autonomy of the technocrats to implement macro-economic reform is apparent while sectoral reforms intended to be carried out by line-ministries depend on which interests should be taken care of (Schmidt 1996). The situation in Indonesia is similar to the Thai example, with an upgrading of the rural sector, but still the main purpose is to channel resources from agriculture to industrial production.

The centrality of the state with regard to capacity and will in these four countries is compared in Table 2.2 by looking at central government expenditure on social programmes and defence. While there is traditionally one social welfare department in the ASEAN-countries, or at least an agency, it typically has a modest budget and does not have a basic-needs mandate. Not surprisingly, during the 18 years from 1972 to 1990, the share of central government expenditure on 'housing, amenities, social security and welfare' among the ASEAN-4 ranged from a 0.9 per cent increase in Malaysia to a decrease of 1.2 per cent in Thailand, a 0.6 per cent increase in Indonesia, and a 2 per cent drop in the Philippines. Placed in the context of Third World development this is markedly lower than the 16.2 per cent average for middle-income LDCs (less developed countries) in 1985 and also much lower than the average in most of the Latin American and South Asian countries. Together with the decrease in defence expenditures, except in the Philippines, this explains partly one 'obvious comparative advantage' *vis-à-vis* other middle-income LDCs.[22] On the other hand, these figures indicate a potential conflict over rising living expectations and pressure on the delivery of public goods and services.

Improvement in the opportunities for the poor to accumulate additional assets is another way to redistribute growth. However, as Table 2.2 shows, public investment in education and health actually dropped significantly in Malaysia, while there was a slight increase in Thailand, Indonesia, and the Philippines.

When considering the general situation of Southeast Asia, the major societal problems are not only those of labour absorption in Indonesia, Thailand and the Philippines but are connected to labour displacement as

Table 2.2 Central government expenditure in the ASEAN-4

	Percentage of total expenditure									
	Defence		Education		Health		Housing, amenities, social security and welfare		Economic services	
	1972	1990	1972	1990	1972	1990	1972	1990	1972	1990
Malaysia	18.5	8.6	23.4	16.1*	6.8	2.6*	4.4	5.5	14.2	20.6
Thailand	20.2	17.3	19.9	20.1	3.7	6.8	7.0	5.8	25.6	22.1
Indonesia	18.6	8.0	7.4	8.4	1.4	2.0	0.9	1.5	30.5	27.6
Philippines	10.9	11.0	16.3	16.9	3.2	4.1	4.3	2.3	17.6	23.6

Source: *Sustaining Rapid Development in Asia and the Pacific*. Development in Practice Series. World. Bank. 1993 and Sixth Malaysia Plan 1991–95 (*) *Economic Planning Unit (EPU)*, Prime Minister's Office, Kuala Lumpur, 1992.

well. Even within the region's established and highly protected manufacturing base the prospects are those of redundancy, continuing foreign control and a growing number of unemployed, volatile and educated youth flocking to cities. Indonesia's labour force grows by 1.6 million annually. The Philippines and Thailand must each find 800,000 new jobs every year, at a time when investors want higher value-added efficiency and productivity from each worker – an effect belying the usefulness of foreign investment in mopping up unemployment.

Rather than a conclusion – drawing up some perspectives

There are different opinions about the prospects discussed above, but in a sociological perspective, the seemingly relatively peaceful social change taking place in Southeast Asia could prove to be of a temporary nature.[23] Given the huge socio-economic divide between the many in the working class and the very few in the elite, it might be anticipated that a class-based (urban worker) revolution or reformist mobilization could occur in one or more of the countries. In the foreseeable future, however, this is unlikely to happen for three reasons. First, the fact that there is some upward social mobility lends hope to poor families that one or more of their children will somehow be able to rise on the socio-economic ladder to middle-class status. Second, an expanded middle class provides a sort of buffer between the two classes at the extremities of the social spectrum. Third, a successful revolution normally requires support from the majority of the population, and in all of these countries the vast majority consists of rural-based families engaged in agriculture or fishing. With their own agenda, these rural folk generally have yet to identify their interests with those of urban workers and vice versa.

   In addition, it should be recalled that it is only in situations of crisis that reformist change occurs. In Southeast Asia there has been neither state capacity nor will to implement a land reform as was the case in South Korea and Taiwan. On the contrary, social unrest has been characterized by demands from the rural areas for agrarian reform; in Indonesia (1965), in Malaysia (1965–68), Thailand (1973–76), and the Philippines on several occasions. Thailand and Indonesia have had regime structures that to varying degrees allowed minor redistributive changes. Consequently, this was done as a deliberate attempt to alleviate social tensions. Reforms aimed at increasing social welfare in the Philippines were resisted by a feudal landlord-class and big business while Malaysia introduced significant reforms creating new employment opportunities.

No country in Southeast Asia except perhaps Malaysia, Singapore and Vietnam, has a binding and properly functioning social contract in the Rousseau sense. The legitimacy of the governments and state bureaucracies in Thailand and Indonesia, and to a certain extent also in the Philippines, is based on economic growth and attempts to promote social order. What is quite unique in Malaysia is the fact that the Bumiputra content of the New Economic Policy and Vision 2020 could be interpreted as a social contract providing interest intermediation. The Bumiputra policy, successful or not in economic terms, provides a strong incentive to force the Malay population into a social compact with the governing UMNO party and the state apparatus. Perhaps, some would suggest that the monarchy in Thailand and the Pancasila ideology in Indonesia perform similar functions. But these examples are extremely fragile because these institutions are based on a personality-cult and thereby create uncertainty with regard to the question of succession. To draw a comparison with the Philippines, which resembles the situation in Latin America, it is clear that, if the legitimacy of the state is based on economic growth, when a personalized social contract breaks down, the result is chaotic instability and lack of popular trust in even the basic functions and performance of the state. Apart from the Philippines, it has not been necessary in the other countries to enter into the bargaining process and shift alliances with various social sectors as is characteristic of the polical process in Latin America.

Capitalism in the late 1980s and 1990s has been marked by social differentiation, not only involving the expansion of capitalist and working classes, but also the emergence of a middle stratum. Together with the increasing segmentation of these classes, new social groups, such as environmental and consumer organizations, human rights and gender-based movements, and, to a certain extent, semi-independent business associations, have achieved considerable popular response. While most public and private labour organizations have been institutionalized and incorporated into the structures of the state (Hewison and Rodan 1994: 253–5), the 1990s have been marked by increasing pressures on the welfare functions of the states in the ASEAN-4. This trend has not yet found appropriate political expression.

What is apparent in this contemporary paternalistic Mandarin-type state model is that uncontrolled capitalism at one level requires governance and agreement on economic policy aimed at negotiating social turbulence to function in the short-run and to become sustainable in the long-term. Legitimacy in the Southeast Asian nation-building strategy derives from one factor only: economic growth. But without welfare or some institutionalized binding compact which satisfies the growing expectations of the middle classes and the working classes, it can be predicted that the Southeast Asian prototype of capitalism will potentially deteriorate into social chaos or revolutionary turmoil. In this respect Southeast Asia is a region susceptible to the movements of capital accumulation at the global level. The autonomy of the custodian state is rather more limited by fluctuations of external influences and up- and down-turns of the world market than was the case for South Korea and Taiwan.

1. For helpful comments and advice I wish to thank Garry Rodan and Jacques Hersh.

2. The gist of the idea about the developmental state as an autonomous and insulated institution was originally displayed in the writings of Alexander Hamilton and Friedrich List and reintroduced by Johnson (1982). A comprehensive review of theories related to the concept is provided by Evans (1989: 1241–1304) especially section 4.3. For a tentative application to the Southeast Asian context, see Chee and Navaratnam (1992: 365–404).

3. For contributions in a neo-institutionalist perspective, see MacIntyre and Jayasuriya (1992) and Doner (1991).

4. Seminal attempts were made in Haggard (1990) and Yoshihara (1988).

5. For an interesting account, see Dixon (1991).

6. For this kind of argumentation see Higgott and Robison (1985) and McVey (1992).

7. However, I will not discuss in detail the political manifestations of these emerging social forces.

8. See also Limqueco et al. (1989: 142–4).

9. There is no theoretical reason why higher growth should lead to higher welfare and the link is not well established. Nevertheless, it is useful to draw a descriptive map of the situation. See Evans (1989: 1292).

10. On the role of labour see Limqueco *et al* (1989), and on the role of peasants, Karunan (1984).

11. See Komin (1991), Lambert (1993) and for a somehow different situation in the Philippines Carino (1991).

12. See for example the ASEAN case-studies presented in Kurth (1989).

13. See in particular the contributions in McVey (op. cit.).

14. This formidable effort of organizing trade unions and farmers' associations ironically resembles what most mainstream political scientists today call 'civil society'. I therefore agree with the position taken by Hewison and Rodan that 'the current deepening of civil society in many parts of Southeast Asia is not a new phenomenon and does not represent an evolutionary transition from authoritarianism to democracy'. See Hewison and Rodan (1994).

15. For various reasons and explanations see Hewison and Rodan (1994).

16. For a systematic account of evidence on personalized relationships between state officials, be it technocrats, the military or bureaucrats on the one hand, or representatives of TNCs, multilateral organizations and core states on the other, see Kowalewski (1989: 81–5).

17. Jones (1988). Referred to by Dixon (1991: 215–16).

18. The following draws on Bautista (1992: 34–5).

19. Note that 12 countries are included in the sample. Kowalewski (1989: 73).

20. The following builds on Crone (1993: 61–4).

21. For a comprehensive analysis and critical evaluation of this reform, see Charoensin-olarn (1988: 203–57).

22. However, lately a new arms race is taking place partly because of the withdrawal of the United States from the region but also caused by fear of a potentially strong and self-confident China.

23. The following relies partly on Thomas (1991: 7).

Bibliography

Alagandram, S. and **I. bin H. A. Rahim** (1984) *The Right to Strike and Lockout*, ILO/UNDP/ASEAN Programme of Industrial Relations for Development, Geneva

Amsden, Alice (1989) *Asia's Next Giant: South Korea and Late Industrialization*, Oxford University Press, New York

Bautista, Romeo M. (1992) *Development Policy in East Asia. Economic Growth and Poverty Alleviation*, ISEAS, Singapore

Bhalla, Surjit and **Homi Karas** (1992) 'Growth and Equity in Malaysia: Policies and Consequences', in Teh Hoe Yoke and Goh Kim Leng (eds) *Malaysia's Economic Vision. Issues and Challenges*, Pelanduk Publications, Petaling Jaya, Malaysia

Carino, Ledfivina V. (1991) 'Social Dimensions of Philippine Industrialization', *Regional Development Dialogue*, **12**(1), Spring

Castells, Manuel (1991) 'Guest Editor's Introduction', in Manuel Castells (ed.) 'Transnational Corporations, Industrialization, and Social Restructuring in the ASEAN Region', *Regional Development Dialogue*, **12**(1), Spring

Charoensin-olarn, Chairat (1988) *Understanding Postwar Reformism in Thailand*, Duang Kramol, Bangkok

Chee, Stephen and Ramon V. Navaratnam (1992) 'The Role of the Public Sector in Economic Growth', in Teh Hoe Yoke and Goh Kim Leng (eds) *Malaysias's Economic Vision. Issues and Challenges*, Pelanduk Publications, Petaling Jaya, Malaysia

Crone, Donald K. (1988) 'State, Social Elites, and Government Capacity in Southeast Asia', *World Politics*, vxl(2)

Crone, Donald K. (1993) States, Elites and Social Welfare in Southeast Asia, *World Development,* **21**(1)

Crouch, Harold and **James W. Morley** (1993) 'The Dynamics of Political Change', in James W. Morley (ed.) *Driven by Growth. Political change in the Asia-Pacific Region,* M. E. Sharpe, New York

Dixon, Chris (1991) *South East Asia in the World-Economy,* Cambridge University Press, New York

Djojohadikusumo, Sumitro (1989) *Indonesian Economic Developments During Four Five-Year Plans 1969/1970–1988/89,* Centre for Political Studies, Jakarta

Doner, Richard F. (1991) *Driving a Bargain: Automobile Industrialization and Japanese Firms in Southeast Asia,* University of California Press, Berkeley

Evans, David (1989) 'Alternative Perspectives on Trade and Development', in Hollis Chenery and T. N. Srinivasan (eds) *Handbook of Development Economics,* **2**, North Holland, Amsterdam

Haggard, Stephan (1990) *Pathways From the Periphery: The Politics of Growth in the Newly Industrializing Countries,* Cornell University Press, Ithaca, New York

Hewison, Kewin and **Garry Rodan** (1994) 'The Decline of the Left in Southeast Asia', *Socialist Register 1994,* Merlin Press, London

Higgott, Richard and **Richard Robison** (eds) (1985) *Southeast Asia. Essays in the Political Economy of Structural Change,* Routledge and Kegan Paul, London

Jayasuriya, Sisira (1992) 'Structural Adjustment and Economic Performance in the Philippines', in Kanishka Jayasuriya and Andrew MacIntyre (eds) *The Dynamics of Economic Policy Reform in South-East Asia and the South-West Pacific,* Oxford University Press, Singapore

Johnson, Chalmers (1982) *MITI and the Japanese Miracle,* Stanford University Press, California

Jones, Gavin W. (1988) 'Urbanization Trends in Southeast Asia: Some Issues for Policy', *Journal of South East Asian Studies,* **19**(1)

Karunan, Victor P. (1984) *A History of Peasant Movements in Thailand and the Philippines,* Plough Publications, Hong Kong

Komin, Suntaree (1991) 'Social Dimensions of Industrialization in Thailand', *Regional Development Dialogue,* **12**(1), Spring

Kowalewski, David (1989) 'Dependence, Development and State Repression', in George A. Lopez and Michael Stohl (eds) *Contributions in Political Science,* No. 209. Greenwood Press, New York

Kurth, Helmuth (1989) *Economic Growth and Income Distribution,* Moed Press, Manila

Lambert, Rob (1993) *Authoritarian State Unionism in New Order Indonesia*, Working Paper No. 25 Asia Research Centre, Murdoch University

Lamberte, Mario (1992) *Philippine External Finance, Domestic Resource Mobilization and Development in the 1970s and 1980s*, PIDS and ISS, The Hague

Limqueco, Peter, Bruce McFarlane and **Jan Odhnoff** (1989) *Labour and Industry in ASEAN*, Journal of Contemporary Asia Publishers, Manila and Wollongong

Lindsey, Charles (1992) 'The Political Economy of International Economic Policy Reform in the Philippines: Continuity and Restoration', in Kanishka Jayasuriya and Andrew MacIntyre (eds) *The Dynamics of Economic Policy Reform in South-East Asia and the South-West Pacific*, Oxford University Press, Singapore

McFarlane, Bruce (1988) 'Growth and Cycles in Southeast Asian Development', *Journal of Contemporary Asia*, **18**(2)

MacIntyre, Andrew J. and **Kanishka Jayasuriya** (eds) (1992) *The Dynamics of Economic Policy Reform in South-East Asia and the South-West Pacific*, Oxford University Press, Singapore

Martin, Kurt (1991) 'Modern Development Theory', in Kurt Martin (ed.) *Strategies of Economic Development. Readings in the Political Economy of Industrialization*, Macmillan (in association with Institute of Social Studies), Basingstoke

McVey, Ruth (ed.) (1992) *Southeast Asian Capitalists*, Cornell Southeast Asia Program, Ithaca, New York

Pernia, Ernesto M. (1991) *Some Aspects of Urbanization and the Environment in Southeast Asia*, Report No. 54, Asian Development Bank, Manila

Ratanakomut, Somchai et al. (1994) 'Urban Poverty in Thailand: Critical Issues and Policy Measures', *Asian Development Review*, **12**(1)

Robison, Richard (1986) *Indonesia: The Rise of Capital*, Allen & Unwin, Sydney

Schmidt, Johannes Dragsbaek (1993a) *ASEAN i en forandret International Arbejdsdeling* (ASEAN in a Changing International Division of Labour), Den Ny Verden, Centre for Development Research (CDR), Copenhagen, **26**(2)

Schmidt, Johannes Dragsbaek (1993b) *In the Shadow of the Pacific Century – Comparative Perspectives on Externalities' Influence on Economic Policy-Making in Southeast Asian Would-be NICs*, (Department of Development and Planning, Aalborg University), Development Research Series, No. 31, Aalborg

Schmidt, Johannes Dragsbaek (1994) *Increasing Exports in a Decreasing World Market: The Role of Developmental States in the ASEAN-4*,

(Department of Development and Planning, Aalborg University), Development Research Series, No. 33, July, Aalborg

Schmidt, Johannes Dragsbaek (1996) 'Paternalism and Planning in Thailand: Facilitating Growth Without Social Benefits', in Michael Parnwell (ed.) *Uneven Development in Thailand*, Avebury, Aldershot

Thomas, M. Ladd (1991) 'Social Changes and Problems Emanating from Industrialization in the ASEAN Region', in Manuel Castells (ed.) 'Transnational Corporations, Industrialization, and Social Restructuring in the ASEAN Region', *Regional Development Dialogue*, **12**(1), Spring

UNDP (1993) *Human Development Report*, United Nations Development Programme (UNDP), New York

World Bank (1993) *Sustaining Rapid Development*, World Bank, Washington D.C

Yoshihara, Kunio (1988) *The Rise of Ersatz Capitalism in South-East Asia*, Oxford University Press, Singapore

Yuen, Ng Chee, Sueo Sudo and **Donald K. Crone** (1992) 'The Strategic Dimension of the "East Asian Developmental States"', *ASEAN Economic Bulletin*, **9**(2)

CHAPTER 3

The emergence of the middle classes in Southeast Asia and the Indonesian case

Richard Robison

It was only two decades ago, in the 1960s and 1970s, that popular Western images of Southeast Asia were dominated by rice fields and peasants, military coups and generals. In many ways this was not an unreasonable representation of the social and political realities of the times. However, it is some measure of the rapidity with which changes have taken place in the intervening period that these old icons have been replaced almost entirely by new ones in which factory workers and businessmen, politicians and traffic jams are the central images, at least in Australia, which is the most sensitive barometer of the Asian region in the so-called West.

These new images are part of a wider changed perception of Southeast Asia from that of a poor region requiring the assistance of the West in its efforts to develop to that of an industrial hothouse which offers boundless new opportunities for trade and investment but also constitutes a potential economic threat to jobs and businesses in the West. Given the increasing rate of relocation of industrial production from countries like Australia to Southeast Asia and the corresponding influx of Southeast Asian low-wage manufactures and investment, particularly into property markets, such anxieties have concrete resonances.

At the heart of these new Western images of Southeast Asia is a burgeoning middle class. In a reaction of almost cargo cult proportions, the emergence of this new social force has been greeted in advanced industrial economies as a vast new consumer market at a time when recession and low growth rates have depressed business prospects at home. Western and Japanese companies jostle to capture the new demand for cars, computers, processed foods, educational services, films and TV soaps.

But Western images of the new middle classes in Asia extend beyond this crass vision of new consumers. For Western conservatives, the new middle classes of Asia are models of hard work and sacrifice, in contrast to what they see as moral decline and decay in the West. The views of Lee Kuan Yew, with their emphasis upon frugality, discipline and conformity, represent the ideals of the stability and order that, for them, characterizes middle-class society. In the middle classes of Asia, Western conservatives recapture the lost frontier of Western capitalism and its values.

However, Western liberals derive quite different messages from the new middle classes of Asia. In the pluralist and liberal orthodoxies, the bourgeoisie and middle classes are regarded as the bearers of modernity and of 'rational' culture that emphasizes the maximization of the interests of individuals. As such, they are the natural enemy of political and social systems based on inherited privilege, arbitrary power and particularist and personalist social and political relationships. By definition they are the champions of free markets, liberal reforms and democratic politics (Lipset 1959; Hagen 1971: 73–82; Hoselitz 1971: 183–92). For liberals it was the middle classes who took to the streets in Manila and Bangkok to topple authoritarian regimes and usher in democratic reform just as they attempted to do in Tianenmen Square. It is the new middle classes who are seen to lead new movements for human rights and environmentalism, to be producing innovative films and demanding and producing an increasingly vital and incisive press. In other words, they see the middle classes of Asia performing the same historical role they were perceived to have undertaken, in liberal mythology, in the West.

In reality, both views are right and both are wrong. Just as they did in Europe, the middle classes of Asia have supported a variety of cultural, social and political philosophies and action. In Singapore, perhaps the most middle class of all societies, it is a primarily middle-class constituency that supports a hierarchal and conformist social and political order of truly Orwellian proportions. Support for the organic statist ideals and corporatist institutions of President Suharto's New Order is substantial among Indonesia's middle class. At the same time the leadership of fundamentalist religious movements in Malaysia, Indonesia and the Philippines can be categorized as predominantly middle class. At the other end of the spectrum the increasing number of progressive movements pushing for reforms in the political, legal and administrative systems, urging accountability of the state and its officials, greater civil rights and freedoms and rule of law are also essentially middle-class movements.

Yet the apparent contradictions involved here do not indicate a massive case of social schizophrenia or that social identity and action are random, accidental, coincidental or chaotic. Rather, they derive partly from the fact that the middle class is a vast and internally differentiated social category with differing sets of interests and relationships with other social and political forces. Bunched together within this unwieldy category are highly educated and wealthy urban professionals, powerful public and private sector managers, the new technocrats, and the huge armies of poorly paid clerks, teachers, and salesmen and women. These blend at the top into the world of corporate capital and, at the lower end of the spectrum, into the world of the petty bourgeoisie and the working class.

As well, the historical role of these complex elements fluctuates with the changing configuration of political and social power and alliances. In Indonesia, a powerful state has played the central role in the process of capitalist industrialization and it is in the shadow of this dominating apparatus that the various elements of the middle classes have been required to consider their social and political options and form their alliances. In contrast to this authoritarian corporatist environment, the middle classes of the Philippines are located within an oligarchic system of powerful families whose wealth was originally constructed around the ownership of large commercial haciendas. In other cases, elements of the middle classes have allied themselves with declining sections of the petty bourgeoisie in political and social projects heavily influenced by ideologies of radical or reactionary populism. Hence the involvement of school teachers, clerics, minor officials and intellectuals in the Communist movements throughout Southeast Asia in the 1950s and 1960s. The impact of the middle classes cannot be understood outside the context of these sets of historically specific conjunctures.

What is the middle class?

How can the middle classes be fitted into the process of political and social struggle that characterizes capitalist development? As Burris has observed, the major political cleavages of capitalist society usually cut right through the centre of the middle class, as understood in terms of a white-collar group (Burris 1986: 344–45). To address this apparent paradox, various commentators, including Poulantzas, Wright, Carchedi, Mill and Ehrenreich, divided the middle class (which they identified in terms of Weberian notions of the distribution of market rewards) into subcategories based on a technical division of labour: the supervisors and supervised, the possessors of knowledge and the semi-skilled. These subdivisions essentially identified control and supervisory capacity exerted over labour rather than income and status. Within such a schema, the routine clerical workers were separated from the middle classes and included with the proletariat (Burris 1987: 33). We must also assume that the foremen are separated from the workers and included with the bourgeoisie.

The differences in the political behaviour of various elements of the middle class do not, however, simply reflect different objective interests. These may be identical. What may differ are the sets of political opportunities and potential alliances that are available for their achievement.

In both liberal and Marxist paradigms the middle classes represent a force for liberal reform: the subordination of the state to the individual and to civil society through legal guarantees of civil rights and political systems of representation. But this picture is less a universal truth than a representation of the circumstances of social and political revolution in eighteenth and nineteenth century Britain and, to a lesser extent, France and the Netherlands. In this situation, absolutist states represented the last stand of exclusivist privilege, restraining the development of bourgeois society. As a range of theorists have observed, the political and social circumstances within which industrial capitalism has emerged are varied, and political outcomes and the role of various classes themselves vary with the timing of the industrialization process and with the configurations of state and class power.[1] In Bismarkian Prussia, pre-Czarist Russia, Meiji Japan and, more recently, in Korea and Taiwan, the state has been the incubator of capitalism, not the last bastion of feudal privilege. Consequently, social and political alliances have taken a different form. Not surprisingly, the links between the middle classes and the state have been close.

A large part of the problem for both liberals (including neo-Weberian pluralists) and Marxists is the tendency to perceive the middle classes (and, indeed, all classes) as conscious and cohesive agents of social and political change. In some circumstances this may be the case. However, the middle classes are better understood, not as agents of change but as products of sets of social and economic dynamics that require change. For example, reforms which make accountable the state and its officials are better understood as guarantees of predictability for investment and property than as capitulation to the haranguing of middle-class reformers. Clearly middle classes benefit from an efficient and accountable state apparatus but they are rarely its progenitors.

The emergence of the middle class in Southeast Asia

The economies of Southeast Asia have, with the exception of the Philippines, grown at a substantial rate of around 6 per cent per annum over the past decade. Their combined economies have grown from around US$150 billion in 1980 to just under US$400 billion in 1992. This growth is part of a process of capitalist industrialization which has also involved critical structural changes. As indicated in Table 3.1 there has been a uniform and dramatic decline in agriculture's share of GDP over the same period. In Indonesia this fell from around 45 percent to just

Table 3.1 Changing structures of production (percentage of GDP)

	Indonesia		Malaysia		Singapore		Thailand		The Philippines	
	1970	1992	1970	1992	1970	1992	1970	1992	1970	1992
Agriculture[1]	45	19	29	16	2	0	26	12	30	22
Non-manufacturing industry[2]	9	19	13	15	10	10	9	11	7	9
Manufacturing industry	10	21	12	29	20	28	16	28	25	24
Services, etc.	36	40	46	40	68	62	49	49	39	45

Sources: The World Bank, *World Development Report* series, Bank Negara Malaysia, *Annual Report 1992.*
[1] Includes forestry, animal husbandry and fisheries.
[2] Covers mining and quarrying, construction, electricity, gas, water.

Table 3.2 Economically active population by type of occupation (percent)

	Indonesia		Malaysia		Singapore		Thailand		The Philippines	
	1980	1990	1970	1988	1980	1991	1984	1988	1985	1990
Professional, executive and technical	2.9	3.7	6.0	7.4	11.7	16.9	3.1	3.2	5.6	5.7
Administrative and managerial	0.1	0.2	1.4	2.1	6.3	8.8	1.2	1.3	1.0	1.1
Clerical, sales and service workers	20.7	23.4	27.6	33.0	28.4	28.4	13.4	15.1	25.6	24.8
Agricultural, animal husbandry, fishermen and hunters	55.8	49.9	35.0	30.6	n.a.	0.2	70.0	64.4	49.1	41.0
Production-related workers, transport equipment operators, labourers and others	20.5	22.8	30.0	26.9	53.6	45.7	12.3	16.0	18.7	27.4

Sources: Euromonitor Plc., *International Marketing Data and Statistics*, various issues, Central Bureau of Statistics, comparison of Indonesian Population Censuses, Jakarta.

under 20 percent. At the same time, the shares of the manufacturing and services sectors rose substantially across most of the countries.

Not suprisingly, these shifts in the structure of economy are reflected in the occupational structure of the societies (see Table 3.2). Sustained declines in agricultural workers are balanced by increases in the numbers in the manufacturing and service sectors. In the countries where manufacturing for export based on low-wage labour is in full swing, increases in the number of workers in manufacture have been dramatic. In Indonesia, for example, between 1971 and 1990 the numbers of workers in this category increased from 2.7 million to 4.4 million (Hadiz 1993: 191). On the other hand, where manufacturing has become more technology-intensive, and where standards of living have risen most, as is the case particularly in Singapore and Malaysia, it is in the service sector that the numbers have most dramatically increased.

The new industrial and service corporations and factories required their new armies of professionals, managerial and technical experts, and those categories grew substantially in most countries. Between 1980 and 1990 they increased from 3.0 per cent to 3.9 per cent in Indonesia. If we crudely extrapolated these figures to percentages of the total population we would arrive at a figure of around 7 to 8 million for Indonesia's professional, managerial and technical middle class. In the same categories, Malaysia increased from 7.4 per cent to 9.5 per cent, Singapore from 18.0 per cent to 25.7 per cent, Thailand from 4.3 per cent to 4.5 per cent and the Philippines from 6.6 per cent to 6.8 per cent.

The very large category of clerical, sales and service workers is a difficult one to translate into social categories. The huge armies of minor clerks, sales persons and service workers straddle the lower middle-class, the petty bourgeoisie and the working class. Nevertheless, in terms of such indicators as ideology, educational background, concern for the ascendancy of legal frameworks and hierarchal systems of meritocracy, accountability of the state and its officials as well as guarantees of citizenship rights, a large number of these would, by most definitions, fall into the middle class category. In particular, as graduate unemployment takes a firmer grip, educated individuals will increasingly be found in minor clerical and sales occupations and will join the ranks of unemployed or underemployed. The political implications of this are fairly clear and it is interesting to note that the leadership of populist, fundamentalist movements in the Middle East and North Africa have drawn leadership from these elements of the middle class.

In general there is evidence that the process of economic growth and industrialization has produced improvements in absolute living standards. A range of indicators suggest that these improvements in living standards are widespread. Ownership of cars, telephones and televisions has increased, there are more physicians per head of population, safe wate

Table 3.3 Changing patterns of consumer expenditure (per cent shares)

	Indonesia		Malaysia		Singapore		Thailand		The Philippines	
	1980	1991	1986	1991	1980	1992	1980	1991	1980	1992
Food, drink and tobacco	69.3	53.4	45.7	39.1	29.7	26.3	46.7	33.6	57.3	56.5
Clothing and footwear	5.1	5.6	6.5	6.9	9.8	8.3	10.8	12.7	6.5	7.5
Housing and fuels[1]	12.2	15.2	8.1	11.8	9.8	12.0	9.5	8.1	17.4	16.1
Household goods and services	7.3	8.0	8.4	11.3	10.2	10.4	3.2	8.0	7.8	10.5
Health, leisure and education	2.2	3.5	7.2	16.9	17.1	22.4	15.7	18.8	7.5	4.8
Transport and others	3.9	14.3	24.1	14.0	23.4	20.6	14.1	18.8	3.5	4.6

Source: Euromonitor Plc., *International Marketing Data and Statistics,* various issues.
[1] 1982 for Thailand, instead of 1980.

Table 3.4 Indicators of relative welfare

	Year	Indonesia	Malaysia	Singapore	Thailand	The Philippines
Population with access to safe water (%)	1975–80	11.0	n.a.	n.a.	25.0	n.a.
	1988–91	51.0	n.a.	n.a.	76.0	n.a.
People per physician	1980	26,820.0	4,190.0	1,370.0	8,290.0	9,270.0
	1990	7,030.0	1,180.0	820.0	4,360.0	8,120.0
People per telephone	1980	300.4	22.9	3.4	94.0	68.8
	1990–92	142.9	9.3	2.6	32.3	58.8
People per television	1980	48.8	12.2	6.0	57.7	48.3
	1990	16.7	6.8	2.6	8.8	20.8
People per car (passengers and commercial) in use	1980	113.1	15.3	9.7	52.7	56.6
	1990	64.7	11.2	6.4	24.9	49.0
Average household consumption of electricity (terajoules)	1980	0.2	1.9	7.2	1.2	1.4
	1990	0.7	4.1	12.1	2.4	0.9
Advertising expenditure as % of GDP	1982	0.17	0.47	0.79	0.32	0.35
	1991	0.39	0.81	0.92	0.67	0.45

Sources: Euromonitor Plc., *International Marketing Data and Statistics,* various issues. The World Bank, *World Development Report* series. UNDP, *Human Development Report* series.

supplies are more accessible. People are being required to spend less of their incomes to meet the requirements of basic needs such as food and clothing (see Table 3.3 and Table 3.4).

The debate over concentration of wealth and whether there is an increasing gap between the rich and the poor is in full swing.[2] The macro-data available suggest that concentration of wealth in the hands of the top echelons of the population is substantial but, on the surface, not any more so than in older industrial capitalist societies such as the US. The share of income or consumption expenditure for the top 10 per cent ranges from 27.9 per cent in Indonesia to 37.9 per cent in Malaysia (Table 3.5) but the reliability of data for the former, particularly relating to income from business profits and political rents is highly dubious. From a political perspective, the real problems lie not so much in the concentrations themselves but in the rapid changes of situation and in the frustrations of those who become economically or politically marginalized in the process.

Concentration of wealth is of considerable relevance for the sort of political role the middle classes are likely to play. In societies where the substantial wealth of the nation is in the hands of small elites, middle classes are likely to be small, relatively powerless and apprehensive of the radi-

Table 3.5 Percentage shares of income or consumption

	Year	Lowest 20%	Highest 20%	Highest 10%
Indonesia	1976[1]	6.6	49.4	34.0
	1990[1]	8.7	42.3	27.9
Malaysia	1973[2]	3.5	56.1	39.8
	1989[2]	4.6	53.7	37.9
Singapore	1982–83	5.1	48.9	33.5
Thailand	1975–76[4]	5.6	49.8	34.1
	1988[4]	6.1	50.7	35.3
The Philippines	1970–71[1]	5.2	54.0	38.5
	1988[1]	6.5	47.8	32.1
United States	1978[3]	4.6	50.3	33.4
	1985[3]	4.7	41.9	25.0

Sources: The World Bank, *World Development Report,* 1994 and 1984 editions.
[1] Data refer to expenditure shares by fractiles of persons, ranked by per capita expenditure.
[2] Data refer to income shares by fractiles of persons, ranked by per capita income.
[3] Data refer to income shares by fractiles of household, ranked by household income.
[4] Data refer to expenditure shares by fractiles of persons, ranked by per capita income.

cal and populist potential of large impoverished masses. They tend to ally themselves closely with the elites. Where the profile is flatter, middle classes tend to be larger and more independent of the state and the elites. Most important, where the working classes and the rural masses are drawn more fully into the economy on a more equitable basis the middle classes are less fearful of radicalism from those quarters and therefore more likely to take reformist political stands.

The political impact of the Southeast Asian middle class

As Westerners look with increasing apprehension (or hope) at the forms of capitalism unfolding in Asia, the middle class becomes, in Western eyes, a strategic factor in the calculation. Will the burgeoning middle classes be a force for liberal reform? Will they constitute the non-commissioned officers of authoritarian hierarchies? Will they constitute a set of restraints upon the booty capitalism that appears to be entrenching itself and the alliances of state officials and robber barons that are so prevalent?

While it has been from the ranks of the urban middle class in Southeast Asia that its liberal and democratic reformers have been drawn, the middle classes have also provided, in other circumstances, the bulwark of support for authoritarian developmentalist regimes and the leadership for populist movements. In general, like middle classes everywhere, their primary concern has been with the establishment of a system of market and career rewards based on credentials and qualifications. Generally it is when governments have obstructed this through incompetence in economic management or gross corruption and rent-seeking that middle-class resentments have been activated. Translating these resentments into action is, however, usually dependent upon seizing opportunities provided by fractures within the state and structural pressures for reform, or constrained when there is a threat of social revolution from the working and rural classes. Response from the middle classes has been governed both by the prevailing configuration of social and political power and by the differing situations of the various elements of the middle classes: the urban professional and managerial middle classes and the lower-level clerical and sales categories which often respond in different ways and form different social alliances.

The Philippines represents, in some respects, a unique case in Southeast Asia. Social power has been in the hands of a ruling class originally built upon ownership of large agricultural estates. Its political system is characterized by a weak state apparatus and a form of client–patron oligarchy focused upon Congress. The involvement of the middle classes in popular struggles against attempts by Marcos and the military to impose a more centralized and authoritarian form of state power in the mid-1980s was widely hailed by Western observers as evidence of a natural affinity between democratic reform and middle-class interest. We must be careful with this interpretation for two reasons. First, an important factor in the unpopularity of the Marcos regime was its inability to deliver economic growth and rising living standards. Had it achieved this, the middle classes might have supported the Marcos regime against the challenges from the contending alliances of powerful families that eventually triumphed. Second, the political outcome was not a genuinely representative form of democracy with guarantees and protections of civil rights so much as a reinstatement of the authority of the major families within the mechanism of an oligarchic form of electoralism. The middle classes proved to be a secondary player in a game dominated by the bourgeoisie and the various institutional interests within the state apparatus (Anderson 1988: 3–31; Hutchison 1993: 191–212).

A more usual pattern of middle-class political activity in the region has been that of alliance with modernizing authoritarian states. Perhaps an extreme example of this is to be found in Singapore where the state and the ruling PAP (People's Action Party) have placed economic growth, rising living standards and political stability at the centre of their claims to legitimacy. In a political system where the divisions between the state, the government and the PAP are blurred, the middle class has a real stake in PAP rule. Not only is the PAP a party that nurtures the growth of industrial capitalism and that has presided over an unprecedented rise in living standards, it also embodies the ideals of middle-class interest: hierarchy and privilege built upon meritocracy; the importance of credentials and education; and the ascendancy of laws. It is the middle class that has benefited most from sustained economic growth. Middle-class Singaporeans now enjoy an equivalent access, not only to the typical range of consumer goods enjoyed by middle classes in advanced industrial countries, but to such commodities as foreign holidays, high-quality housing and overseas education (Rodan forthcoming).

It is also significant that there was no powerful domestic bourgeoisie to play a central role in the political history of Singapore, at least in the 1950s, 1960s and even into the 1970s. In the early standoff between the PAP and labour, the interests of the middle class were clear.

There are clearly irritations between the middle classes and the state in Singapore. Reaction to overregulation and the government's obsessive concern for conformity has come to the surface in rising rates of emigration and protests against the more bizarre examples of social engineering,[3] as well as in the government's reluctance to allow emerging middle-class organizations freedom to operate and criticize outside the institutions set up by the state. Despite these problems, the Singapore state can in important ways be regarded as essentially a middle-class authoritarianism.

Until recent years, the configuration of power in Malaysia has rested, in the political sphere, with a Malay-dominated state apparatus and, in the economic sphere, with a Chinese bourgeoisie. The rapid growth of a Malay white-collar class has been consciously facilitated by the state, not only as a byproduct of its success in providing the conditions for sustained economic growth but as a result of specific policies and legislative measures introduced to ensure that ethnic Malays entered the ranks of the professional and managerial middle classes in large numbers. Without the intervention of the state, establishing quotas in such areas as employment and bank loans, the strong surge in the Malay middle class would not have been possible. While, as Crouch (1993: 133–8) points out, elements of the middle classes have been amongst the government's most trenchant critics in such areas as human rights and the environment, and some have played leading roles in opposition parties such as the DAP and PAS, in the context of a society where economic power had long rested with a non-Malay bourgeoisie, the alliance between the ruling UMNO Party and the upper levels of the Malay middle class is strong.

In Thailand, the burgeoning middle classes have been popularly associated in the West with the riots and demonstrations of 1992 that led to the fall of General Suchinda and his replacement by a democratic coalition under Chuan Leekpai in a development seen as a watershed victory for democracy over military control. Kevin Hewison presents an intriguing rebuttal of many of the assumptions that tied the middle classes and other reformist forces so closely to the 'democratic' changes.[4]

In Hewison's view, the Thai middle class, increasingly independent of the state for employment and favour and no longer needing the state to protect it against Communist revolution began, during the 1980s, to feel able to move in a reformist direction. However, its primary objectives were not uniformly and specifically to secure democratic government so much as to achieve competent and clean government. This had been demonstrated in the past by the fact that the middle classes had been prepared to support military intervention when it was carried out against civilian parliamentary governments widely perceived to be corrupt and incompetent. Nor was the apparent democratization of 1992, in Anderson's view, an embodiment of middle-class political ideals. It was essentially a transition to electoralism as a mechanism for bourgeois ascendancy over the centralized bureaucratic apparatus rather than constituting any real representative system with attendant liberal guarantees or social justice for the bulk of the population. In a situation where the bourgeoisie, including the large bankers, no longer needed the more coercive aspects of state power to secure their social dominance, the new institutions of electoralism were to allow direct political control by a range of political entrepreneurs representing the interests of the new bourgeoisie in both Bangkok and, increasingly, in the regions (Anderson 1988).

Amongst the middle classes, according to Hewison, the new democracy evoked a widespread fear of the 'dark forces', including criminal elements, competing for control of central government ministries that would enable them to dominate the 'pork barrel' politics resulting from the reforms. In this instance, there is a divergence of views between reformist elements of the middle classes and the new bourgeois political entrepreneurs about the nature and role of the new system of electoralism. For the latter, ideas of representation and legal protections of the rights of citizens sit uneasily with the capture of the mechanisms of parliament and elections for the purposes of dividing spoils. The same factors that saw previous alliances of elements of the military and the middle classes to restore clean and orderly government have not been entirely eliminated. Whether such alliances emerge again depends to a large extent on whether the new political structures can be captured and used by reformist forces to achieve real representative reforms and the establishment of mechanisms of political accountability.

In some ways, Indonesia presents a similar picture in that it is a new bourgeoisie, nurtured within the cocoon of the New Order, that is seeking to secure its political framework. As Suharto's New Order regime developed in the years following its rise to power in 1966 new political institutions began to emerge outside the military structures that initially sustained it. An extensive set of corporatist organizations became the basis of political control and mobilization. These included a state political party, Golkar, and a range of state-endorsed corporatist organizations representing 'functionalist' groups and interests in society such as youth, women, civil servants, business and so on. As a framework for these institutions a complex ideology of organic corporatism was refined, carrying the picture of society as an organic entity in which component parts worked with common purpose towards the common good. The function of the state in this view was to protect the common interest against the emergence of vested interests or destructive and dysfunctional elements. Within such a perception of politics and society there is no place for legitimate opposition as a loyal and constructive force (Robison 1993: 41–7; Reeve 1992: 151–76).

Within the state apparatus itself, the authority and influence of the military was being gradually eroded by the institutions surrounding the Presidency, notably in the 1980s by the State Secretariat. The economic power of the military, particularly in the critical area of procurement was gradually appropriated by the State Secretariat and by Habibie's Ministry of Research and Technology (Robison 1993: 48–9). Powerful elements within the military began to take a critical line towards the Presidency and the business alliances that surrounded it as its growing power began to diminish their own institutional base.

This resentment was exacerbated as the Presidency became the hub around which powerful bureaucratic families were to build a base of social power in the world of corporate capital. Using a variety of market monopolies, including import licences, distributorships for the state oil company, Pertamina, contracts for public works and infrastructure, entry into former state monopoly areas such as television, and generous state bank credit, the Suharto family established Indonesia's most extensive group of *pribumi* corporate conglomerates. Other families with a base in bureaucratic office or with family links to the Presidency were to build similar, although smaller empires. These included the Habibie, Djojohadikusumo, Harmoko and Sutowo families. Below these, in terms of access, were other *pribumi* groups that had been able to gain a foothold, mainly as contractors and constructors under the patronage of the State Secretariat and, in an earlier era, of Sutowo when he ruled Pertamina. These were also sustained by extensive corporate links with both state enterprises and those of the major Chinese corporate conglomerates.[5]

The coming political succession has activated a struggle to determine the form of the post-Suharto political regime. From the point of view of Suharto and the new families and notables it is essential that the political succession does not also involve a transition either to some form of liberal, representative government or to some form of populist military rule. A struggle for power between these incipient oligarchies and their allies, and those elements of the bureaucracy representing, or claiming to represent, a tradition of accountable and regularized authoritarianism simmers just beneath the surface. Suharto's strategy has been to root out hard-line elements from the military and, at the same time, to open avenues within the corporatist structures of the state, offering opportunities for careers and political access to the new notables and middle classes. Within Golkar the ascendancy of the military and the state bureaucracy is being challenged by the new families and other emerging notables. In an ironic twist, the military hard-liners, formerly ultra-authoritarian, have become advocates of democratic reform and social justice, opposing vehemently the rise of the new families surrounding the Presidency.[6]

Where then, does the emerging middle class fit into all of this? Just as in Thailand, the dramatic growth of the private sector has increased the influence and autonomy of professional, managerial and technical sections of the urban middle classes. Employment is increasingly located outside the state, often in international companies, giving a greater degree of independence from the state. Combined with a period of greater political openness in the past five years, individuals drawn from the ranks of middle-class professionals have provided a very public and sophisticated critique of the nature and exercise of political and economic power in Indonesia, focusing in particular upon the position of the families and conglomerates in the corporate world and on the political impotence of parliament and parties within the corporatist system.[7]

However, unlike Thailand, there is no structure of electoralism which allows government to be captured by alliances of business and middle classes bypassing the system of centralized bureaucratic control. More important, the bourgeoisie remains closely tied to the state and its strata of officials who still continue to provide the important avenues to business success through various rent-seeking practices. Because the private sector is primarily Chinese and thereby constrained in playing a prominent public political role, the bourgeoisie in Indonesia is somewhat distorted. Nevertheless, even those non-Chinese sectors of the bourgeoisie, able to engage in public politics, support the existing corporatist structures, particularly Golkar, and it would be unwise to assume that had the Indonesian bourgeoisie been entirely indigenous things would have been different.[8] The problem lies in the general relationship between state and capital rather than in communal factors. Whatever the case, the reality is that the bourgeoisie has been neither a bridgehead for a political transition towards electoralism nor a potential ally in such an enterprise for the middle class.

For the bulk of the upper, urban middle class, sustained rises in living standards, manifest in better housing and increasing disposable incomes, together with the attendant culture of consumerism, have led to a general support for the New Order, despite irritations with aspects of the arbitrary exercise of authority by military and civil bureaucrats and increasing contempt for the highly public corporate depredations of the first family. Most look to reform within the system. The government-sponsored ICMI (the Association of Muslim Intellectuals), recently established by Suharto and Habibie, has been quite effective in providing a channel for Muslim middle-class professionals to reach positions of influence within the state apparatus and provides a clear indication of the President's intention to build a larger civilian base to the regime (Hefner 1993: 1–35; *Forum Keadilan* 1993).

While the upper middle classes have been the focus of considerable study, there has been a real dearth of systematic work on the politics of the huge numbers of lower-middle-class clerks and officials, in the regions as well as the cities. Although they are benefiting from the general process of economic growth, there are signs that the strong traditions of populism continue to be influential among this sector of society. The student movement is one interesting indicator of this. Two distinct elements appear to have emerged. On the one hand is a 'yuppie' element increasingly concerned with careers and achieving the material rewards of life, paralleling the growing conservatism of upper-middle-class students in the West. On the other hand, whereas in the 1960s and 1970s the radical and reformist elements of student politics were led by upper-middle-class students, with the expectation that some sort of reform would spontaneously emerge from within the system as a result of their moral critiques, radical student movements in the past decade have increasingly been led by lower-middle-class students and have engaged in alliances with peasants and disadvantaged social groups, invoking populist social critiques.[9] Growing graduate unemployment, especially amongst those from the lesser universities and schools, may allow this student radicalism to persist beyond university. The volatile and enthusiastic support for the PDI in public parades and demonstrations preceding the 1992 election, which gave it a reputation of being the party of the young and politically progressive, clearly embodied important elements drawn from a frustrated, lower-middle-class youth.

While this undoubted potential for populist radicalism among the lower middle classes is currently unable to find the institutional channels for political outlet, the post-Suharto era may provide a much more fluid environment.

Conclusion

Clearly the middle classes of Southeast Asia have had an important impact on the political life of the region in the past decade. At particular junctures, when regimes have faltered, in Thailand in 1992, in the Philippines in 1985, and earlier, in Jakarta in 1965, the middle classes have burst onto the streets and briefly appeared to dominate the political scene. However, they have always done so in alliance with other groups and following political dynamics set in train by the disintegration of regimes or pressures for change which result from structural shifts in the economy. Nor have the democratic systems that have emerged always contained the liberal elements associated with middle-class interests. Human rights and rule of law as well as guarantees of individual freedoms are often excluded from the constitutions and practices of the new systems of electoralism.

The tendency to view the middle classes primarily as political agents perhaps misses their real political significance as the products of the set of social and economic dynamics that present political powerholders with new sets of constraints, contradictions and choices. Let me give some examples from recent developments in Indonesia. In Indonesia, middle-class liberals have long pressed for freedom of the press. While the Indonesian government has never systematically sought to impose the *Pravda*-like controls that are evident in Singapore, the Ministry of Information has the power to issue and revoke licences and has been prepared to close papers when limits have been transgressed. As recently as June 1994 it has exercised this power with the banning of three leading papers; *Tempo, Detik* and *Editor*. However, it is becoming more difficult for the Ministry to control the press through the blunt instrument of the ban. This is not because of political pressure from a liberal middle class wanting freedom of expression. Such pressures can be easily ignored. One factor is that the Indonesian press is increasingly big business. Profits depend on advertising revenues which, in turn, derive from capturing large slices of a market that has come to demand interesting and entertaining news. It is a market increasingly able to gain access to media internationally. One of the President's children, Bambang, owns satellites which are dependent for profit on renting time to media outlets like CNN. The middle class, therefore, exercise their power as the new mass consumers, the bourgeoisie as investors trying to gain an edge on competitors in this increasingly lucrative market.

A second example is the banking industry. State banks in Indonesia have long been used as milch cows for favoured investors. Recent scandals have exposed massive levels of bad debt flowing from loans made to investors which often appear to have contravened normal banking practices and involved figures close to the centres of power.[10] For decades, middle-class critics have been highly critical of these rent-seeking activities but without any result. While observers are still sceptical that dramatic and sustained reforms will be made, bank officials and business figures involved in the latest scandal have been tried and imprisoned. The factors that have led to this action have arisen from a variety of structural economic constraints on rent-seeking behaviour. Decline in oil prices and increasing reliance on foreign loans and aid for budget revenues have led to greater competition for relatively shrinking fiscal resources. Pumping state funds into the state banking system to cover corrupt practices will be a practice increasingly difficult to maintain. In addition, the increasing integration into international financial systems for both private and public investors in Indonesia requires greater adherence to international practices and standards to maintain credibility. It is becoming more difficult to ignore the pressures for regularization and transparency of procedures when Indonesian public companies (such as Telekom and Indosat) seek listing on the New York Stock Exchange. In other words, middle-class objectives relating to the accountability of the state and its officials may be achieved, not through the agency of middle-class political action, but as a result of the structural pressures of an industrializing capitalist economy.

Finally, a reorganization of Indonesian politics is being undertaken in the context of the coming succession from Suharto. In a regime where the institutions of authority and all the career channels were to be found in the civil and military bureaucracies, the Presidency has begun to open new channels through Golkar and ICMI, giving careers and political opportunities to middle-class political entrepreneurs. Ironically, these new institutions do not constitute a shift to representative government but a means of retaining authoritarian corporatism. It is a process of civilianization without democratization. Most important, the shift does not represent a response to middle-class pressure but a strategy by the Presidency to nurture new political constituencies and institutions to enable the newly entrenched capitalist oligarchies to maintain power into the post-Suharto era in the face of potential threats from nationalist populism within the military and liberalism from elements of the upper middle classes. The middle classes have gained a political place but one within the structures of authoritarian corporatism, not liberal democracy, and to suit the agendas of capitalist oligarchies.

While this analysis suggests that the political and ideological identity of the middle classes is relative to the prevailing configuration of political and social power and dependent upon potential alliances and coalitions, there is nevertheless something universal about the middle-class interest in its requirement for frameworks of predictable and institutionalized regulation and law that enshrines routine and credentials and protects against the arbitrary exercise of power. With the rapid internationalization of finance, information, education, technology and management, elite elements of the middle class are able to some extent to escape the confines of their national environments within new, transnational, technical institutions. Here, their incomes, status, authority, ideology and, indeed, their language achieve a degree of insulation and autonomy. Their interests can be secured within a variety of political and social regimes.

1.   The classical exposition of this approach was Moore (1966). See also Kurth (1979).

2.   An interesting and detailed recent article focusing on rural villages in East Java suggests that a pattern of increasing polarization between rich and poor does not apply in that case. See Edmundson (1994: 133–48).

3.   The most recent examples are moves by the government to end any support for single mothers and to enable parents to take legal action against children who refuse to support them adequately in old age. Although these are moves made on the grounds of protecting Asian values of family life they are clearly directed against the encroachment of the welfare state.

4.   Hewison (forthcoming). In his study, Hewison draws on the following: Laothamatas (1992); Phongpaichit (1992); Anderson (1990).

5.   Robison (1986: chapters 7–10); Hwan (1989: chapters 6–10) an interesting look at some of the children of officials entering the world of business is to be found in *Prospek* (1993).

6.   There is now a huge literature on this. For an idea of the sort of democratic populism being purveyed see retired General Soemitro's statement in *Detik* (1994).

7.   See for example Sjahrir (1993); Kwik (1993); Pamungkas (1993); Sukardi (1993); Wahid (1993).

8.   The recent influx of new notables, including figures from the powerful new bureaucratic-business families and retired officials as well as an assortment of *nomenklaturas* from the corporatist institutions of the New Order is examined in *Tempo* (1993). It includes the move of members of the Suharto family onto the party's central supervisory board.

9.   See Pamuntjak (1993 (see especially pp. 37–40 for details of the social and economic origins of student leaders)).

10.   There is an enormous literature on this. See *Tempo* (1994a) and *Tempo* (1994b).

Bibliography

Anderson, Benedict (1988) 'Cacique Democracy in the Philippines: Origins and Dreams', *New Left Review*, No. 169

Anderson, Benedict (1990) 'Murder and Progress in Modern Siam', *New Left Review*, No. 181

Burris, V. (1986) 'The Discovery of the New Middle Class', *Theory and Society*, **15**(3)

Burris, V. (1987) 'Class Structure and Political Ideology', *Insurgent Sociologist*, **14**(2)

Crouch, Harold (1993) 'Malaysia: Neither Authoritarian nor Democratic', in Kewin Hewison, Richard Robison and Garry Rodan (eds) *Southeast Asia in the 1990s: Authoritarianism, Democracy and Capitalism*, Allen Unwin, Sydney

Detik, 5–11 January (1994)

Edmundson, Wade C. (1994) 'Do the Rich Get Richer, Do the Poor Get Poorer? East Java, Two Decades, Three Villages, 46 People', *Bulletin of Indonesian Economic Studies*, **30**(2)

Forum Keadilan, 16 September (1993)

Hadiz, Vedi (1993) 'Workers and Working Class Politics in the 1990s' in David Collier (ed.), *The New Indonesia Assessment 1993*, Department of Political and Social Change, Australian National University, Canberra, Monograph No. 20

Hagen, Everett E. (1971) 'How Economic Growth Begins: A Theory of Social Change', in Jason L. Finkle and Richard W. Gable (eds) *Political Development and Social Change*, John Wiley and Sons, New York

Hefner, Robert W. (1993) 'Islam, State, and Civil Society: ICMI and the Struggle for the Indonesian Middle Class', *Indonesia*, 56, October

Hewison, Kewin (forthcoming) 'Emerging Social Forces in Thailand: New Political and Economic Roles' in Richard Robison and David S. G. Goodman (eds) *The New Rich in Asia: Mobile Phones, McDonalds and Middle Class Revolution*, Routledge, London and New York

Hoselitz, Bert F. (1971) 'Economic Growth and Development: Non-economic Factors in Economic Development', in Jason L. Finkle and Richard W. Gable (eds) *Political Development and Social Change*, John Wiley and Sons, New York

Hutchison, Jane (1993) 'Class and State Power in the Philippines' in Kewin Hewison, Richard Robison and Garry Rodan (eds) *Southeast Asia in the 1990s: Authoritarianism, Democracy and Capitalism*, Allen and Unwin, Sydney

Hwan, Shin Yoon (1989) *Demystifying the Capitalist State: Political Patronage, Bureaucratic Interests, and Capitalists in Formation in Soeharto's Indonesia*, Ph.D. thesis, Yale University

Kurth, James (1979) 'Industrial Change and Political Change: A European Perspective' in David Collier (ed.) *The New Authoritarianism in Latin America*, Princeton University Press, Princeton, NJ

Kwik, Kian Gie (1993) *Indonesia Observer*, 29 November

Laothamatas, Anek (1992) *Sleeping Giant Awakes? The Middle Class in Thai Politics*, Paper presented to the Conference on Democratic Experiences in Southeast Asian Countries, 7–8 December 1992, Thammasat University-Rangsit, Bangkok

Lipset, S. M. (1959) 'Some Social Requisites of Democracy', *American Political Science Review*, **53**, September

Moore, Barrington (1966) *The Social Origins of Dictatorship and Democracy: Lord and Peasant in the Making of the Modern World*, Beacon, Boston

Pamungkas, Sri Bintang (1993) *Forum Keadilan*, 9 December

Pamuntjak, Laksmi (1993) *The Indonesian Student Movement in the 1980s and 1990s: The Development of Resistance by a Marginalised Minority*, unpublished honours thesis, Murdoch University

Phongpaichit, Pasuk (1992) *The Thai Middle Class and the Military: Social Perspectives in the Aftermath of May 1992*, Paper presented to the Annual Conference of the Thai Studies Group, October 1992, Australian National University, Canberra

Prospek, 6 March (1993)

Reeve, David (1992) 'The Corporatist State: The Case of Golkar', in Arief Budiman (ed.), *State and Civil Society in Indonesia*, Monograph No. 22, Monash University Centre of Southeast Asia Studies, Clayton, Australia

Robison, Richard (1986) *Indonesia: The Rise of Capital*, Allen and Unwin, Sydney

Robison, Richard (1993) 'Indonesia: Tensions in State and Regime', in Kewin Hewison, Richard Robison and Garry Rodan (eds) *Southeast Asia in the 1990s: Authoritarianism, Democracy and Capitalism*, Allen and Unwin, Sydney

Rodan, Garry (forthcoming) 'Class Transformations and Political Tensions in Singapore's Development', in Richard Robison and David S. G. Goodman (eds) *The New Rich in Asia: Mobile Phones, McDonalds and Middle Class Revolution*, Routledge, London

Sjahrir (1993) *Warta Ekonomi*, 25 January

Sukardi, Laksmana 1993 *Kompas*, 11 July

Tempo, 9 October (1993)

Tempo, 19 February 1994(a)

Tempo, 26 February 1994(b)

Wahid, Abdurrahman (1993) *Detik*, 4–10 August

Part 2 Classes, culture and political legitimation

Some careful writers maintain that one hour's work a day would suffice, but perhaps this estimate does not take sufficient account of Asia.

Bertrand Russell, 1935

CHAPTER 4

Class, culture and Malaysian modernity

Joel S. Kahn

This chapter[1] offers some remarks on the theme of emerging classes and growing inequalities in Southeast Asia. The main materials presented here are derived mostly from ongoing research in Malaysia. Research on rapidly changing social constellations in Asia raises a large number of theoretical issues, particularly for those accustomed to treating places such as Malaysia as peripheral in one way or another to the world capitalist economy and the global system of nation states. Here I shall refer briefly to some of these issues, although clearly their resolution would require a good deal more than that. Finally, the sketch of the new constellation of economic, political and social forces this contribution offers is based largely on an analysis of developments in Malaysia over the past couple of decades, although many of the processes at work in Malaysia will also be seen to be operating elsewhere in the region. Nonetheless, this contribution will focus on the Malaysian case, and leave it to readers to deal with the question of the extent to which Malaysian processes are being repeated elsewhere.

Malaysia: neomodernization?

Beginning in the late 1960s, Malaysia was set on a relatively new developmental trajectory with major social, cultural, economic and political implications for what had, until then, been a largely colonial social formation. A sociological correlate of this process has been a significant transformation in the patterns of social stratification that prevailed in the colonial and early postcolonial period. Perhaps most frequently alluded to has been the emergence of a (racialized and gendered) industrial proletariat, a consequence of the acceleration of public and private sector industrialization in the years after 1970; the specific character of new capitalist work processes in the new, labour-intensive, export-oriented manufacturing sector; and the ethnic 'restructuring' aims of the Malaysian state.

At the same time there has been a substantial increase in the relative size of what sociologists are wont to call the middle class. Depending on the interpretation of census data, for example, the size of the 'substantial and prosperous' middle class was as high as 24 per cent of the work force in 1980 (Crouch 1985: 31–2). The class grew in significance in the 1980s, so that, using the same calculation, Saravanamuttu (1989) estimates that by 1986, 37.2 per cent of workers were in middle-class occupations. And doubtless the 1990 census will show continued growth in both the absolute and relative size of the Malaysian middle class.

Equally important have been changes at both the very top and bottom of the social hierarchy. On the one hand, recent years have seen the emergence of new kinds of business elites, differing from the traditional capitalist class in the kinds of close relations they have established with political party apparatuses, particularly that of the leadership of the politically dominant United Malays National Organization (cf. Gomez 1990); in the rise of non-Chinese, particularly Malay, business figures in their own right (rather than mere front men for traditional Chinese capitalists);[2] and in the new kinds of corporate activities, and forms of business organization that began to appear in the 1990s.[3]

On the other hand, we can speak of the growth of what we might at this stage call new underclasses in both urban and rural areas. Particularly significant are the large and growing numbers of immigrant (often illegal immigrant) workers from Indonesia, Thailand, Myanmar and elsewhere who work for relatively low wages and often in extremely insecure positions in land clearing and plantation agriculture, construction, domestic service and the so-called urban informal sector.

It is quite clear that making some kind of sense of these stratificational realities calls for new kinds of sociological paradigm. Classically-inspired accounts of the sociological dimensions of modernization – whether liberal or Marxist – have proved inadequate to the task. It is unnecessary here to enumerate once again the faults of classical Marxist schemas of capitalist development, and of the equally problematic liberal accounts of 'modernization', a task in any case adequately performed first by so-called dependency, world systems and neo-Marxist theorists, and, subsequently, by the various approaches glossed by terms such as postmodern, poststructural, and postcolonial. Despite their faults, these various theoretical streams have made it difficult if not impossible merely to resurrect classical narratives of modernization in whatever guise.

At the same time it cannot be said that, while suggestive of new approaches, any of these revisions of the classical modernist metanarratives are proving much more successful at making sense of the processes of social transformation in places like Malaysia than their predecessors have been. For example the various forms of neodependency theory, notably theories of the so-called New International Division of Labour (NIDL), might have appeared to have been vindicated by the recession of the early to mid-1980s in the TNC-dominated, export-oriented economies of places like Malaysia. Indeed at the end of the period a number of such analyses appeared, apparently revelling in the fall of places like Malaysia as proof of the pertinence of their theories (see, for example, Jomo 1987; Bello and Rosenfeld 1990; Mehmet 1986; and Yoshihara 1988). But NIDL theorists

have proved unable to account for the earlier-than-average economic recovery and subsequent rapid economic growth, to say nothing of the
complex, rather than merely bipolar, system of class relationships that
seems to have accompanied these changes. Nor have those who have employed a concept of 'state capitalism' in the Malaysian context provided
us with a very nuanced understanding of either economic changes or political processes and conflicts, particularly in the last seven or eight years.

And yet rapid economic growth, the accelerated capitalist transformation of the Malaysian economy, the rapid 'rationalization' of bureaucratic structures,[4] moves over the last decade to 'privatize' or corporatize
the economy,[5] the rise of party capitalism,[6] and the renewed commitment
to modernizing discourses, at least at the top,[7] make the various 'post-'
characterizations – postmodern, postcolonial, post-Fordist – of the new
Asian societies, however plausible they may be in the American context,
equally problematic. In a real sense we are witnessing, in places like
Singapore and Malaysia, a reworking of the modern project in socially
and culturally distinctive contexts, something that might lead us to label
these processes neo- rather than postmodern.

Before returning to the changing patterns of class relations and inequality, it is necessary to examine this broader, neomodern, trajectory of
social transformation. Of course the most commonly cited indicators of
change are the figures on economic growth. The Malaysian economy has
grown rapidly over a relatively long period of time – average annual
growth rate over the last 25 years stands at 6.5 per cent. That slowed in
the first half of the 1980s to 5.4 per cent per year between 1980 and 1985.
But since then it has again risen to reach 9.7 per cent in 1990, 8.7 per cent
in 1991 and 8.5 per cent in 1992.

Not surprisingly, this growth is associated with a sectoral shift in the
economy, notably a shift from primary production to manufacturing. In
Malaysia, the manufacturing sector grew at a rate of 6.1 per cent a year for
the period 1980–85, 14.1 per cent annually between 1985 and 1990; 15.7
per cent in 1990, 13.9 per cent in 1991, and 13.0 per cent in 1992.[8] While the
percentage of the work force employed in agriculture, forestry, hunting
and fishing fell from 31.2 to 26.0 between 1982 and 1990, the percentage
in manufacturing rose from 15.5 to almost 20 in the same period.[9] Using
a mix of measurements of the relative size of the industrial sector, McGee
has maintained that Malaysia became an NIC (Newly Industrialized
Country) as long ago as 1980 (McGee 1986: 38).

To understand the current situation, it is important to look more closely
at causes of variations in the growth rate in the years since 1970. Probably
the most astute analysis of Malaysia after the economic downturn has
been provided by Khoo Kay Jin who explains the initial successes of the
1970s in the following terms. A combination of economic circumstances,
particularly in the latter half of the seventies, gave the state sufficient
room for manoeuvre when faced with the reluctance of say, domestic capital, to play along or when confronted with demands from below.
Generally favourable international circumstances, including the commodities boom and the worldwide reorganization of capital investment,
state-led demand and investment primarily via public enterprises – and

the discovery of off-shore petroleum – resulted in an exemplary growth rate. Real GDP averaged 7.6 per cent for the 1970–80 period, and 8.6 per cent for the 1975–80 period. Per capita income increased from M$1,100 in 1970 to M$3,700 in 1980. Given that the country was awash with cash including, importantly, petroleum revenues, this impressive performance was achieved without undue stress on the budget and on the balance of payments (Khoo Kay Jin 1992: 50f).

But the situation in the 1980s was quite different. In the early 1980s and again in 1985 the Malaysian economy went into a relatively severe recession. Per capita income actually fell in both 1985 and 1986; state expenditure contracted; unemployment rose from 5 per cent in 1982 to 8.6 per cent in 1987; small and large businesses went bankrupt; the 'urban middle classes experienced declining real incomes'; private consumption declined, leading to a collapse of the commercial sector; rubber prices fell; manufacturing contracted with resultant job losses and, partly as a consequence of the recession in the West, foreign investment began to dry up (Khoo Kay Jin 1992: 51–7). The period was also characterized by a steady rise in state debt and a balance of payments deficit. By the end of 1987, for example, the total accumulated loss of non-financial public enterprises (NFPEs) reached US $850 million, and the external debt service ratio of these increased from 4.6 per cent in 1985 to 5.8 per cent in 1986, and 5.9 per cent in 1987 (see Toh Kin Woon 1990: 21).

In short, while in the first half of the 1980s the predictions of the pessimists about the fragility of Malaysia's developmental trajectory seemed to be coming true, the signs are that Malaysia has come out of the recession of the 1980s very strongly indeed, probably more strongly than most Western economies. As our figures show, high growth resumed from around 1987, unemployment again fell (so much so that in recent years there is much talk in Malaysia of severe labour shortages), poverty rates again began to fall, and incomes have risen steadily.[10]

Class formation or cultural differentiation?

However, statistical measures such as these provide only a very impressionistic picture of the particular ways in which Malaysia's economy is organized. The system of economic institutions that has developed in the years since 1970, and which has generated these indicators of economic growth, has been characterized by three interrelated processes:

- export-oriented industrialization (EOI), particularly in the manufacture of semiconductor components made possible by high levels of direct foreign investment by transnational corporations;
- a rapidly growing state sector – relatively one of the largest in the world – dominated by a relatively large number of financial public enterprises (FPEs), NFPEs and trusts;
- the emergence of a unique system of what we might call 'party capitalism' (often glossed in Malaysia by the term *money politics*).

This transformation was brought about by a realignment of social forces in the years after independence, and, by generating greatly increased levels of national income, itself contributed to significant changes in local patterns of social stratification and class formation.

At the same time there have been significant shifts in the constellation of political and cultural forces. At one level, of course, the Malaysian state and economy are inseparable. Particularly in the years between 1970 and the early 1980s, both the federal and state governments played significant roles either as direct instigators of economic activities through public enterprise, or as important facilitators, particularly for the TNCs, through the provision of infrastructure, tax breaks, regulation of the labour force, and a favourable tax environment in the so-called Free Trade Zones. But to describe Malaysia as a case of 'state capitalism' is somewhat misleading for a number of related reasons, which space forbids me to expand upon here. First, while public sector enterprises, including those involved in manufacturing, are an important component of the economy, there is far from being a state monopoly of economic activities. Indeed over the last year steps have been taken to privatize or corporatize important parts of the public sector itself. Second, the increasing significance of what is referred to here as 'party capitalism' differs in significant ways from a classic system of state or bureaucratic capitalism. Third, although committed to capitalist development, and keen to use the indicators of economic growth to legitimize itself, the Malaysian state often pursues its own political and ideological projects which cannot be explained by imputing a capital logic to the processes of state formation[11] and political conflict. Suffice it here to point out that neomodernization in Malaysia has been a multidimensional and not a unidimensional process.[12] In addition to the economic transformations discussed above, it has involved, among other things, a neomodernist cultural project currently being pushed from the top by the Prime Minister, his Deputy and a circle of powerful politico-business figures close to the current leadership of UMNO (often bringing them into conflict not just with Islamic and other oppositional groups, but with those more dedicated to the continuation of a more traditionalistic, postcolonial project within UMNO itself); the emergence of a more 'authoritarian' political regime, a process most clearly evident in the increasing relative importance of executive powers in the last ten to fifteen years; and a process of social and cultural differentiation at all levels of Malaysian society.[13]

Now, how have these processes of neomodernization affected the patterns of class formation and social stratification which were laid down during the colonial period, and which persisted in only slightly modified form at least until the mid-1960s? At this time two influential conferences, organized by largely middle-class Malays unhappy with their progress under the conditions produced by the independence settlement negotiated between UMNO and the British, set in train significant changes that were to become official government policy some five years later within the framework of the so-called New Economic Policy. And, together with shifts in the global economy, changes introduced through heavy state intervention in the economy under the twin goals of the New Economic Policy (see below) resulted in the emerging institutional complex described above.

The argument developed here is that the main implications of these processes of neomodernization for classes and inequality have been twofold. On the one hand, changes in the years since 1970 attest to a more-or-less steady process of capitalist class formation, in the classical (Marxist) sense of the term. On the other hand, we can witness a second process of accelerated social and cultural differentiation of the population of Malaysia, a process moreover that cannot be explained by or reduced to some underlying process of class formation, but which instead can only be conceptualized within a broader, multidimensional understanding of the project of modernity Malaysia-style. To emphasize the autonomy of this process of differentiation is to take issue with the various recent theories of class formation in which an attempt is made to theorize this differentiation by means of a proliferation of revisionist class concepts, notably the concepts of a new 'middle class', and of an 'underclass'.

It is necessary to begin with a deceptively simple Marxist diagnosis of the social consequences of the developments discussed above, namely that the modernization of Malaysia has generated a pattern of classically capitalist class relations, between a large and growing proletariat, and a new bourgeoisie; in other words a growing polarization of Malaysian society between owners and non-owners of the means of production. Although certain refinements of this approach are clearly called for, particularly taking into account the important distinction between ownership and control, at one level such a diagnosis is largely self-evident. As long as we restrict the meaning of the term 'class' to its classical Marxist formulation, then it is quite clear that the trend over the past few decades has been an increase in the numbers of people who exercise little effective control over the main means of producing subsistence/consumer goods and of generating surplus value. More and more Malaysians have come to depend on wages and salaries to meet even their most basic needs. And increasingly, following Marx, Malaysian economic elites have come to rely on a 'free' market in labour power and constant capital goods in order to supply the inputs necessary to profitable economic activity. Moreover, the decline in the relative size and importance of 'traditional middle-class groups', such as family farmers, small traders, artisans, etc., the self-employed in general, i.e. the traditional petit bourgeoisie, has also contributed to the swelling of the category 'wage and salary earners'. In short, proletarianization in the classical Marxist sense has been probably the dominant tendency in Malaysian class formation over the last twenty to twenty-five years.

However, to argue that proletarianization has been the central characteristic of the transformation of the Malaysian class structure is not to argue that patterns of social differentiation have been in any simple sense bipolar, and hence that *class* allows us in some sense to account for social processes in Malaysia as a whole. To illustrate this point there are two significant aspects of the contemporary Malaysian social landscape that will inevitably escape our grasp if we focus solely on the processes of class formation in the strict sense of the term – namely culture and the middle classes. The problems raised for a simple class analysis by these factors leads us back to a more nuanced understanding of the current situation in Malaysia.

Forgotten elements: culture and the middle classes

The middle classes and the issue of culture, have both been quite system-atically overlooked in general models of Malaysian 'development', par-ticularly those that rely on models of class formation even when these are subjected to substantial doses of revisionist theorizing.

On the face of things there are a number of reasons to look more closely at the so-called middle strata, several of which have already been men-tioned in passing. **First**, this is necessary to fill in a yawning **gap in exist-ing discussions of Malaysian economy and society.** Until now these have focused overwhelmingly either on elites – Chinese capitalists, 'New Malays' in the corporate sector, and political elites (for example, in recent discussions of the UMNO split most attention has focused on the person-alities of the UMNO leaders, Razaleigh, Mahathir and Anwar) – or, es-pecially in NIDL and state capitalism literature, on the 'subalterns' (poor peasants, the urban poor and particular fractions of the working class, es-pecially young Malay women working in the global factories). At the very least, some consideration of the growing middle strata is needed to fill in this picture of social groups in contemporary Malaysia.[14]

To see that this is so let us merely look at the example of the develop-ment of various 'middle strata' in two of the main economic institutions, namely the mainly foreign-owned electronics industries and the complex of state enterprises.

The dominant image of the Malaysian economy in the 1970s has been that of the 'world factory' where the labour-intensive stages of TNC manufacture (particularly of electronics products) is carried out by a large, unskilled, largely female labour force, compelled to work for rela-tively low wages; an absence of unionization and a high rate of labour turnover, together with a lack of significant linkages between the so-called Free Trade Zones and the rest of the national economy. This image was popularized in the flood of writings on Malay women workers and world factories that began to appear in the late 1970s. Here it was main-tained that, at best, countries like Malaysia would be stuck in a low-wage trap, because the Transnational Corporations that dominated the industry were interested in places like Malaysia only as suppliers of cheap labour.

However, partly because of changes in the structure of the world elec-tronics industry and partly because Malaysia has, particularly since the mid-1980s, been attempting to 'lever' its economy upward by selling itself as a high tech/high skill environment, there is evidence that the 'world factory/low wage' model of the export sector is no longer adequate, pre-suming it was in the first place. Singapore, of course, has had considerable success in this area, achieving what is often called a Second Industrial Revolution by selling itself as a 'global city'; encouraging higher-tech industries; pushing low-wage/low-tech investment offshore (e.g. the 'growth triangle' with southern Malaysia and Indonesia), and promoting the outflow of Singapore capital into low-wage economies (most recently China). At the same time, Singapore has worked to ensure that regional headquarters and profitable producer services for these activities are lo-cated in the Republic and has pursued various 'human resources' options to upgrade the skill level of its work force. But there is evidence, too, that both because of a conscious strategy to upgrade the technological level of the export sector, and because the nature of the global electronics industry in particular has changed there have in recent years been significant changes in Malaysia as well. These have resulted in a marked shift away from the earlier model of a low-wage factory – a shift that, were it to take hold, would make it necessary to revise significantly our earlier notions about the pattern of differentiation of the industrial work force in particular.

In a paper published several years ago, for example, two Malaysian researchers documented some potentially significant changes in Malaysia's electronics industry, which is centred largely on the so-called Free Trade Zones on Penang island where Malaysia's first semiconductor assembly plant (owned by America's National Semiconductor) was established in 1971. Kamal Salih and Mei Ling Young[15] suggest that, although in the early days, Penang's semiconductor industry was indeed characterized by low wages, a predominantly young, Malay, female work force, and labour as opposed to capital intensivity, more recently, and particularly since the mid-1980s, there is evidence of the beginnings of a 'second phase' or 'second wave' of development. This is characterized by 'declining direct labour' on the one hand, and 'technology deepening' on the other. The suggestion is that a middle stratum of skilled workers, computer scientists and engineers, and middle-level managers has emerged, located somewhere between the unskilled worker on the shop floor and the, often expatriate, general management.

Trends in wages and the structure of the public sector work force in the 1970s and 1980s also differ from the polarization anticipated for the global factory regime. If the former were, at least in the early years, characterized by relatively low wages and a high proportion of semiskilled and unskilled workers, between 1971 and 1985 the growth rate of public sector employees was highest for group A employees (highest level, managerial) at 16.6 per cent, while group B grew at 10.2 per cent, group C (clerical and skilled) at 13.2 per cent, and group D (unskilled) at only 9 per cent (see Ismail Salleh and H. Osman Rani 1992).

The size and growth of these occupational categories has not been anticipated in the main models of Malaysian economy and society, and yet they make a significant impact on the economic life of contemporary Malaysia. For example these are the primary consumers of a wide range of services: shopping centres, restaurants, theme parks, new housing estates, domestic tourism destinations, the arts, advertising, the media and journalism etc. It is, for example, the consumption requirements of these middle classes that more than anything else shape the new urban landscapes of Kuala Lumpur, Johore Baru, and Penang.

How are we to read this evidence of an increased differentiation even of the labour force in public sector enterprises and global factories? The tendency in much of the literature has been to attempt to conceptualize situations such as these by means of revisionist concepts, particularly here of concepts of a 'new middle class'. Following on from the work of writers like Bourdieu, Giddens, Abercrombie and Urry and others, many have suggested that these new middle classes are to be defined as groups differentiated, in class terms, from the bourgeoisie because they do not own/control the means of production and from the proletariat because of their 'ownership'/'control' of credentials, information and/or cultural capital.[16]

This is not the place for a detailed critique of such approaches.[17] Suffice it here to point out four related flaws in such analyses. First, there is a very real problem in assuming some kind of formal isomorphism between the Marxist ownership/non-ownership of the means of production distinction, and revisionist dichotomies such as 'ownership'/'non-ownership' of credentials, skills or cultural capital. The former generates a logical bipolarity, the latter does not. To attempt to formally define a class on the basis of the latter is certain to generate a misleading conflation of class understood in the Marxist sense, and class in its revisionist sense. Second, and following on from this, revisionist attempts to distinguish classes on the basis of degrees of control of cultural capital or credentials will never succeed in generating discrete classes, only a plethora or matrix of socio-cultural 'strata'. There can, in short, be no new middle class, only new middle classes. Third, once cultural differentiation becomes a part of class analysis, its application must extend beyond the confines of groups that could even be remotely linked to the new middle classes. I have in mind here in particular so-called 'underclasses', whose distinction from the proletariat is itself very often also 'cultural', since the groups we tend to label 'underclasses' are very often defined by their culturally undervalorized and/or marginal status. Finally, there is that other dimension of cultural differentiation, namely power. Unequal power relations can never be considered the formal equivalent of class relations, since power is always best understood as a matrix or set of matrices of unequal relationships. Moreover, it is always a mistake, particularly in conditions of late or neo-modernity to reduce the matrix of power relations to the structure of class relationships as already suggested.

A **second** reason why more knowledge of the nature and role of the middle classes would appear to be particularly important lies in the fact that the members of these middle class groups also have a very significant **impact on political life**. As both leaders and members of political and NGO groupings, their concerns shape the contours of political life in significant, often decisive ways.

For example, we might point to the ways in which the so-called New Economic Policy, proclaimed in 1970, which took as its central aims poverty reduction and economic restructuring (to achieve greater economic balance among the 'races'), while often said to be a direct response to the 1969 'racial riots' in Kuala Lumpur, in fact has its origins in the pressure exerted by middle-level Malay bureaucrats and (to a lesser extent) businessmen at the Bumiputra Congresses of 1965 and 1968 (see Gomez 1990: 6; Shamsul 1986: 190; Khoo 1992). This explains to some extent why in its implementation, and in spite of its twin aims, the NEP has always been more about restructuring than direct action to reduce rural poverty, a criticism offered by many authors (see, for example, Mehmet 1986). Put another way, many of the structural changes in the Malaysian economy, such as the rise of the electronics industry, the proliferation of public enterprises, the development of 'party capitalism', developed in response to the concerns of, and political pressure exerted by middle-class Malays in the decade or so after independence.

More recently, evidence of middle-class political concerns, especially on the part of the Malays, is clearly manifest in the conflicts both within UMNO and between UMNO and the other main Malay parties (PAS and Semangat '46). This is not the place for a detailed discussion of the pressures leading up to the 1987 UMNO split, and the subsequent conflicts between UMNO Baru, Razaleigh's Semangat '46 and PAS. Suffice it to point out that far from being, as some observers have suggested, merely a result of a clash of personalities between prominent UMNO leaders, or a simple conflict over the spoils of money politics (although these have clearly played their part), these events testify to some major political/ideological cleavages within the Malay community. Of particular importance is the split between an authoritarian modernism being pushed by the Mahathir/Anwar faction (with the support of the big players in the corporate arena) and middle-level Malay businessmen who are loath to see the 'traditionalist' policies of UMNO replaced by pure economic rationalism. Middle-class groups – Malay, Chinese and Indians – line up on different sides of this split for specific reasons that need to be understood.[18] Here, contrary to the view that there is an intimate relationship between market rationality and democratization (cf. Mehmet 1986), and that of those who look to the middle classes to carry the torch of political modernism, we have a modernist agenda being pushed from above as it were, and hence a modernism that must perforce be authoritarian at the same time. This makes sense only when the special circumstances of the middle classes are taken into account.

Since the early 1970s, in spite of the declining importance of the Malay nationalist agenda[19] (which has since the colonial period found its most important political constituency among middle-class Malays), the Islamization process has also been closely linked with the concerns of urban, middle-class Malays. This means that an understanding of the sociological dimension of perhaps the most significant element in modern Malaysian political culture is quite simply beyond our reach if we ignore what one local academic described to me as the 'Islamization of the middle class or, what amounts to the same thing, the middle-classization of Islam'.

Finally, since the above discussion has focused largely on the concerns of the Malay middle classes, we might also mention the changing situation of the Chinese business classes, and particularly of the owners of small and medium-sized industries (SMIs). Small Chinese business has probably suffered more than most during the years since 1970 – both because of the pro-Malay policies that flowed from the restructuring aims of the NEP, and because of the preferential treatment increasingly awarded to foreign capital in the export zones. At the same time this group may well have grown as a consequence of the developing linkages with the foreign sector. As a consequence it appears that a party like the DAP has been able to draw support from this group in its opposition to the Barisan government.[20] Once again the middle classes are shaping political conflict, although as often as not in ways that depart significantly from the classical model of a modernizing/democratizing middle class.

The **third** reason why more attention needs to be paid to the middle classes, one already implied in the above, has to do with the question of **legitimacy** at a more general level. As Immanuel Wallerstein and a number of others have pointed out, the legitimacy of modern forms of political domination rests on legitimation in the eyes of the middle strata more than it rests on the legitimized domination of working classes and other 'subaltern' groups. The reverse is probably true as well, namely that challenges to the established political order are more likely to come from a crisis in the establishment of legitimation among the middle classes than from elsewhere. To take just one example – the current challenge to the cultural authority of states posed by the eruption into the public sphere of the previously suppressed 'voices' of subaltern groups – women, ethnic minorities, indigenous peoples and the like – can probably be better explained by reference to the conditions under which intellectuals and other members of the middle classes have been led to 'speak the voice of the subaltern' as it were, rather than to any increased dissatisfaction or even direct 'empowerment' of members of these subaltern groups themselves. Moreover, to the extent to which regime maintenance does require ideological cooptation of subaltern groups, that too is generally achieved through and/or mediated by members of the middle strata, who may in turn challenge regime legitimacy by 'mobilizing' (re-presenting) the voices of the dominated as part of their struggle against political elites. This rather specific role played by the middle classes in both the legitimation of and resistance to regimes of power is clearly evident in the recent political conflicts in Malaysia discussed above, suggesting once again that more attention should now be paid to the middle classes than has been the case in the past.

A fourth reason to bring the middle classes into the picture is closely bound up with the issues of **culture** and **reflexivity** (see Kahn 1989). Briefly stated the argument is as follows. There are two senses in which 'culture' needs to be, to paraphrase Geertz, relocated at the centre rather than on the periphery of our analyses of Malaysian 'development'. Firstly, there is what we might term the hermeneutic argument that analyses of human social life that do not come to grips with the problem of meaning are ultimately without validity. This is at once an epistemological and an ontological argument in that it maintains both that the processes of social analysis are hermeneutic processes (however much we wish to view them as objective) and that those social processes we seek to analyse are themselves meaningful. There is, thus, no major gap between intellectual 'analysis' and social life in general. Put another way, our 'analyses' are part of the processes we seek to represent rather than mere offstage reflections on them. Hence so-called 'structural models' of Malaysian social, economic and political process are incomplete **both** because they presume objectivity on the part of the observer (an absence of a prior relation between subject and object of knowledge) **and** because they ignore the central importance of meaning in all forms of social life in Malaysia.

Culture in a somewhat different sense is of particular importance in the Malaysian context as well, because 'culture' – thought of now in the anthropological sense of cultural differentiation – is a crucial aspect of the meaningful experience of most Malaysians. As we have already seen, for example, political life, the legitimation of political regimes and political conflict, whatever else they involve, almost inevitably involve questions about the content of and boundaries between different 'cultures' – Malay culture, Chinese culture, Indian culture; the relationship between Asian and Western culture; the role played by religion, particularly Islam in Malaysian culture, etc. It is quite simply inadequate to maintain that such a way of conceiving of Malaysian political process, for example, is an epiphenomenon of the economy, that, in other words, what are perhaps too frequently termed 'questions of ethnicity' are mere cloaks for more 'real' or 'material' bases of political action.

We have already seen how 'structural' models of Malaysian political economy fail even in their own terms, at least in part because they ignore the centrality of culture in both senses of the term. The question then is, what is or should be the place of culture within our analysis of the Malaysian social formation? Here let me merely make two observations. When we focus on the central significance of culture (in the second sense) within the culture (in the first sense) of contemporary Malaysia, as expected, we would find in many places where so-called 'identity politics' is now coming to the fore, that the project of culture building, i.e. of speaking for a culture, of drawing the boundaries within and between cultures, of defining the content of 'Malay culture', 'Chinese culture', 'Islamic culture', 'modern culture', 'Asian culture' etc., while it might purport to be about a pre-existing cultural differentiation among 'the masses', is in fact very much a project of at least certain members of the middle classes. Indeed, to the extent that intellectuals are members of the middle classes this is almost true by definition, since the role of intellectuals in modern states has largely been defined by their function as guardians of/spokespersons for national or regional cultures.[21] This suggests, incidentally, that a consideration of Malaysian culture must involve looking at the production and consumption of cultural representations among the middle classes, although it should also, of course, involve looking at the cultural content of all dimensions of social life.

But putting culture back into our analyses of Malaysian development is not nearly straightforward as might at first sight appear. The reason why the classically-conceived anthropological project of cultural analysis is bound to fail has already been suggested when I drew attention to the dual nature of the hermeneutic project, i.e. that it is, and must self-consciously (reflexively) recognize itself as, a project which involves the subject as much as it does the object of knowledge. In other words when we produce a cultural 'interpretation' – say of Malaysian political life – we cannot leave ourselves out of it since we are at least part of the dialogical process, if not the sole author of that interpretation. This, moreover, suggests another intimate link between culture and the middle classes – in so far as we, as observers, are located and locate ourselves within emerging global middle classes, then we stand in a relationship increasingly of identity with middle-class culture builders in Malaysia. This means that accounts of Malaysian middle-class culture are less objectified/externalized than are the kinds of (colonial) anthropological accounts of cultural otherness, where the relationship between observer and observed is largely either ignored, or confused with a dyadic interaction that takes place over a very short space of time in the 'field', with ultimately no genuine commitment on the part of the observer except to his/her academic career 'back home'.

Conclusion

Arguing for the necessity of bringing culture and the middle classes back into the picture of the nature of Malaysian development is, of course, to beg many questions. For example this contribution has so far avoided the question of the meaning of 'middle classes', although by consciously using plural forms of the noun (middle classes, middle strata) I have suggested that the assumption of unity in much of the literature is problematic. This is not to suggest, for example, that doing so will ultimately produce a more accurate – because more complete – picture of Malaysian life. The result of focusing on the middle class gaze will on the one hand provide an additional perspective on political and cultural processes in a place characterized, like many other parts of the world, by a highly fragmented vision. On the other hand, by drawing our attention to the importance of taking account of processes of social and cultural, as well as class differentiation it leads us also to consider the emergence of new kinds of politico-business elites as well as the formation of what some have misleadingly termed a new underclass, thus genuinely giving us a more complete picture of the Malaysian social formation.

In sum, the processes of political, social and cultural differentiation that characterize Malaysian neomodernity must be analysed as separate from the dominant process of class formation, and not be, as in the work of a number of revisionists, subsumed under an increasingly elaborate but ultimately sterile conceptual apparatus of 'new' classes. Only in this way can we hope to provide an understanding of Malaysian modernity that is sensitive to its Malaysia-specific and multidimensional character.

1. The research on which this chapter is based is supported by the Australian Research Council. A large number of people have provided assistance with the work in Malaysia. I would like particularly to thank my co-worker, Maila Stivens, Halim Salleh, Francis Loh, Khoo Kay Jin, K. S. Jomo, Francesco Formosa, Clive Kessler, Norani Othman, Alberto Gomes, Ken Young, Goh Beng Lan, Wendy Mee, Richard Tanter and, especially, Hah Foong Lian who proved a superb research assistant in Malaysia.

2. The so-called 'Melayu Baru' (New Malays). See, for example, any number of discussions such as an article entitled 'The new Tan Sri businessmen' in *The Star*, 6 June 1992. See also Gomez (1990).

3. Gomez (1990) traces the rise of big players in the complex network of overlapping directorates, takeovers, etc. that are associated with the rise of UMNO corporate activities. Others have demonstrated clearly how the kinds of 'traditional' Chinese business activity

that emerged during the colonial period have been transformed in similar ways (cf. Heng 1992; Sieh 1992).

4. See, for example, Ismail Salleh and H. Osman Rani (1991).

5. See, for example, Toh (1989; 1990).

6. See Gomez (1990).

7. See Kahn (forthcoming).

8. Source: *Trends in Developing Economies 1993*, IBRD / World Bank, Washington, D.C.

9. See Department of Statistics, 1990, *Laporan Penyiasatan Tenaga Buruh*. Kuala Lumpur, 1991.

10. Part of the reason for Malaysia's ability to recover relatively quickly from the world recession of the 1980s was the increasing volume of investment from Japan and the rest of East Asia beginning in the second half of the 1980s.

11. In this there are important parallels between Malaysia and Singapore. Of the latter, Castells has argued convincingly that rapid industrialization was certainly facilitated by heavy state involvement, but that the state had its own political and strategic reasons for doing so. See Castells (1991).

12. This critique of the unidimensionality of the existing developmental theories, both Marxist and liberal, is heavily influenced by the work of Johann Arnason. See, for example, Arnason (1987; 1987/8).

13. For discussions of some of these changes see various contributions in Kahn and Loh (eds) (1992); Kahn (forthcoming).

14. As I have also suggested, work on the poor needs also to look at the very significant presence of overseas migrants, especially illegal migrants from Indonesia, Thailand, Indochina and Burma.

15. See Kamal Salih and Mei Ling Young (1989). See also Rajah Rasiah (1991); Fong Chan Onn (1989). The general conclusion that NIDL theory has consistently underestimated the extent to which, especially in Southeast Asia, the electronics industry has been upgraded and will continue to be able to upgrade to higher value-added and higher-waged manufacture is supported by a recent study of TNC operations in Indonesia, Malaysia and Singapore by Ian Chalmers. Chalmers (1991).

16. For such an analysis in the Malaysian context, see Saravanamuttu (1989).

17. For a very good critique of existing analyses of the development of new middle classes, see Barbalet (1980). Barbalet also argues convincingly that we cannot, as some analysts have assumed, merely resurrect Weber's notion of 'status' in contexts such as these, since for Weber social differentiation on the basis of status has its origins in distinctly premodern cultural categories, while if nothing else the cultural distinctions at work in contemporary Malaysia are definitely not traditional.

18. Here I am in substantial agreement with the position taken by Khoo Kay Jin (1992) who demonstrates clearly that Mahathir's authoritarian modernism (my term, not his) is perfectly consistent with his earlier position on the 'Malay dilemma', and that he was never a wholehearted supporter of the premises that led to the NEP. For the more conventional view, see Shamsul (1988).

19. Two writers who document this decline are Muhammad Ikmal Said (n.d.) and Rustam Sani. (Sani (1993)).

20. Ian Chalmers has suggested that it is precisely this group, not a genuine 'national bourgeoisie' because of their links to foreign firms, that appears to be the focus for mounting opposition to Singapore's PAP regime. See Chalmers (1991).

21. See, for example, Bauman (1992), although here he argues that this function is now in decline.

Bibliography

Arnason, J. (1987) 'The Modern Constellation and the Japanese Enigma', Part 1, *Thesis Eleven*, **17**: 4–39

Arnason, J. (1987/88) 'The Modern Constellation and the Japanese Enigma', Part 2, *Thesis Eleven*, **18/19**: 56–84

Barbalet, J. M. (1980) 'Limitations of Class Theory and the Disappearance of Status: the Problem of the New Middle Class', *Sociology*, **20**(4): 557–75

Bauman, Zygmunt (1992) 'Love in Adversity: On the State and the Intellectuals, and the State of the Intellectuals', *Thesis Eleven*, **31**: 81–104

Bello, Walden and Stephanie Rosenfeld (1990) *Dragons in Distress: Asia's Miracle Economies in Crisis*. Institute for Food and Development Policy, San Francisco

Castells, Manuel (1991) *Four Asian Tigers with a Dragon Head: A Comparative Analysis of the State, Economy and Society in the Asian Pacific Rim*, Universidad Autonoma de Madrid, Instituto Universitario de Sociologia de Nuevas Technologias. Working Paper No. 14

Chalmers, Ian (1991) 'International and Regional Integration: The Political Economy of the Electronics Industry in ASEAN', *ASEAN Economic Bulletin*, **8**(2): 194–209

Crouch, Harold (1985) *Economic Change, Social Structure and the Political System in Southeast Asia*, ISEAS, Singapore

Fong Chan Onn (1989) 'Wages and Labour Welfare in the Malaysian Electronics Industry', *Labour and Society*, **14**: 81–102

Gomez, Edmund Terrence (1990) *Politics in Business: UMNO's Corporate Investments*, Forum, Kuala Lumpur

Heng Pek Koon (1992) 'The Chinese Business Elite of Malaysia', in Ruth McVey (ed.) *Southeast Asian Capitalists*, Southeast Asia Program, Cornell University, Ithaca, New York

Ismail Salleh and H. Osman Rani (1991) *The Growth of the Public Sector in Malaysia*, ISIS (Institute of Strategic and International Studies), Kuala Lumpur

Jomo, K. S. (1987) 'Economic crisis and policy responses in Malaysia', in R. Higgot and Richard Robison (eds) *Southeast Asia: Essays in the Political Economy of Structural Change*, Routledge, London

Kahn, Joel S. (1989) 'Culture: Demise or Resurrection', *Critique of Anthropology* **9**(2): 5–26

Kahn, Joel S. (forthcoming) 'Globalism, Tourism, Heritage and the City in Georgetown', in Michel Picard and Robert Wood (eds) *Tourism, Ethnicity and the State in Asian and Pacific Societies*

Kahn, Joel S. and **Loh Kok Wah, Francis** (eds) (1992) *Fragmented Vision: Culture and Politics in Contemporary Malaysia*, Allen & Unwin, Sydney and University of Hawaii Press, Honolulu

Kamal Salih and **Mei Ling Young** (1989) 'Changing conditions of labour in the semiconductor industry in Malaysia', *Labour and Society*, **14**: 59–80

Khoo Kay Jin (1992) 'The Grand Vision: Mahathir and Modernisation', in Joel S. Kahn and Francis Loh Kok Wah (eds) *Fragmented Vision* . . .

McGee, T. G. (1986) 'Joining the Global Assembly Line: Malaysia's Role in the International Semiconductor Industry', in T. G. McGee et. al., *Industrialisation and Labour Force Processes: A Case Study of Peninsular Malaysia*, The Research School of Pacific Studies, Australian National University, Canberra

Mehmet, Ozay (1986) *Development in Malaysia: Poverty, Wealth and Trusteeship*, Croom Helm, London

Muhammad Ikmal Said (n.d.) 'Malay Nationalism and National Identity', Unpublished paper

Rajah Rasiah (1991) 'Reorganization of Production in the Semi-Conductor Industry and Its Impact on Penang's Position in East Asia', in Muhammad Ikmal Said and Johan Saravanamuttu (eds) *Images of Malaysia*, PSSM, Kuala Lumpur

Rustam A. Sani (1993) *Melayu Baru dan Bangsa Malaysia: Tradisi Cendekia dan Krisis Budaya*. Utusan Publications & Distributors, Kuala Lumpur

Saravanamuttu, J. (1989) *Kelas Menengah dalam Politik Malaysia: Tonjolan Perkauman atau Kepentingan Kelas*, Kajian Malaysia, **7**(1&2): 106–126

Shamsul, A. B. (1986) *From British to Bumiputera Rule*, Institute of Southeast Asian Studies, Singapore

Shamsul, A. B. (1988) 'The "Battle Royal": The UMNO Elections of 1987', in Mohammed Ayoob and Ng Shee Yuen (eds) *Southeast Asian Affairs 1988*, Institute of Southeast Asian Studies, Singapore

Sieh Lee Mei Ling (1992) 'The Transformation of Malaysian Business Groups', in Ruth McVey (ed.) *Southeast Asian Capitalists*, Southeast Asia Program, Cornell University, Ithaca, New York

Toh Kin Woon (1989) 'Privatization in Malaysia: Restructuring or Efficiency', *ASEAN Economic Bulletin* **5**(3): 242–250

Toh Kin Woon (1990) 'The Role of the State in Southeast Asia', in Jacques Pelkmans and Norbert Wagner (eds) *Privatization and Deregulation in ASEAN and the EC*, ISEAS, Singapore

United Nations Development Programme (1992) *Human Development Report 1992*, Oxford University Press, Oxford and New York

World Bank (1993) *Trends in Developing Economies 1993*, IBRD/World Bank, Washington D.C.

Yoshihara, Kunio (1988) *The Rise of Ersatz Capitalism in South-East Asia*, Oxford University Press, Singapore

The legitimacy of the public sphere and culture of the new urban middle class in the Philippines

Niels Mulder

In this cultural analysis of ideas and identity formation in the Philippines, the middle class is defined by a relatively advanced level of education. Especially since the 1970s, education has dramatically increased and has contributed to what is termed the contemporary new urban middle class. The emphasis in this chapter will be laid on the differences between generations before and since the 1970s and their relationship to the public sphere. The observations presented stem from a wider investigation about the cultural construction of the public sphere and identity formation in Southeast Asia. It addresses the question of how various strata develop ideas about social life beyond their private – familial and communal – embeddedness. How do these segments perceive wider society? This inquiry complements earlier, more conventional anthropological work on the cultural construction of everyday life in Java, Thailand and the Philippines (Mulder 1992a; 1994a; 1996) and draws on many years of participant observation and research among members of the urban middle class, combined with a systematic analysis of the construction of knowledge in the major newspapers and social studies school texts.

First of all, some observations are presented on the physical, economic, and cultural environments which have come into existence together with the rapid expansion of the middle class over the past twenty-five years. What is the middle-class's cultural perception of the wider world in the midst of dynamic social changes which involve among other things rapid urbanisation, an evolving service sector, consumer culture, and career-focused education? Yet, it would be frivolous to understand the culture of the new urban middle class solely on the basis of its present experience. In order to provide a sufficient analysis, the new urban middle class must be situated in its proper historical context of what is argued as the persistently low legitimacy of the public sphere. This point of departure explains the cultural affliction that is known as 'colonial mentality'. The following analysis delineates the characteristics of the Filipino culture of the public sphere in terms of possibilities for personal identification and it tries to provide an understanding of the 'individualism' or 'self-centredness'. In the context of contemporary urban society there are tendencies which stimulate a retreat into individual, primary group-centred, and particularistic options.

Members of the urban middle class are the producers, disseminators, and consumers of mainstream and alternative ideas; they are the mainstay of 'public opinion', and their milieu is the matrix of ideas about a desirable order of society. In the 1960s and early 1970s, the ideas that were most vociferously expressed were inspired by an optimism about the possibility of social reconstruction; they were about democratization. Expressed by the challenge from students to hierarchical, non-democratic structures, authoritarian decision-making, and neocolonial dependence, such ideas were developed by a generation which was the product of different circumstances from the present. Higher education was still a privilege, and the students' criticisms and violent protests, aiming to achieve a more open society, were more or less directed at the elder members of their own class.

Since then, much has changed. Compared to the rather elitist older group, a quantitatively impressive, new middle stratum has come to the fore whose members are the product of novel conditions that shape their lives and outlook, their culture and political demands. Although the critical ideas of democratization and social change are still present, the groups who express them, and who can on occasion be clearly heard, have been marginalized and are no longer able to dominate the public agenda. Apart from being set within a nation-wide cultural and political history, and recently within a period of martial law and dictatorship, the new middle class is very much the product of a period of important changes in the productive basis of society with its demands for new skills and offer of new lifestyles. This changing environment is considered first.

Modern Environment

Urbanization

Until well into the 1960s, the component towns and cities of what is now known as Metro Manila were still separated from each other by stretches of greenery, *esteros* or marshland; buildings were low, and the skyline inconspicuously marked by churches, tall trees, and rare old three-storeyed buildings. Slums had already come into existence, and the traffic had never really moved smoothly. At that time the place was of manageable proportions. With a population moving towards 3 million inhabitants (Caoili 1988: 107), it was not precisely a village, yet many people in professional circles had the idea that they knew each other.

Thirty years later, all this has changed beyond recognition. The various parts of the metropolitan area have grown together while trees and greenery have disappeared. Not only have the fields of Makati given way to a mini-Manhattan, but high-rise buildings have become the norm everywhere, whether as condominiums for the corporate executives and professionals, international hotels, office towers, or gigantic shopping malls. Most road space is choked up by traffic and its fumes, and there is a permanent shortage of public transport. Slums have spread everywhere, and over 8 million people crowd urban space. Of course, they have not driven out the original inhabitants, but it is important to bear in mind that most are rural immigrants and new to the city with its very complex environment, which appears overwhelming and anonymous to them.

Economy and education

The urban economy attracts people from all over the country who try their luck in the capital; most barely survive, but many fill the slots created by the modern sector. Since the late 1960s, the opportunity structure has changed significantly, and the rapid growth of the service sector – education, commerce, engineering, media, banking, tourism, government planning – has stimulated the demand for sophisticated skills and advanced levels of training, to which the education industry eagerly responded. The result was that in 1990, approximately 1.5 million students were enrolled nationwide at the tertiary level, which means a fivefold increase compared with the mid-1960s. This education-to-learn-a-profession process has given rise to a novel sort of people: the new urban middle class.

This class differs from earlier university-educated generations who went to college as a matter of privilege and were susceptible to ideologies and social analysis; they did not primarily train for a career. This contrasts with the swelling masses that came to the colleges and universities in the early 1970s, for whom professional and career considerations constituted the overwhelming motivation. This outlook coincided with the state administration's policy of economic development and with the long period of repression of free thought, which resulted in the birth of a new generation of politically indifferent and socially unaware students, the so-called martial law babies.

Fully exposed to the propaganda of Marcos, this is a vast generation bred on a curriculum that is precariously low in social science and humanities content. The aim of school education was to form a technocratic mind and an orientation towards national development, that is, an orientation towards the future and away from the past. This brought a generation to the fore that is generally devoid of a sense of history, a generation exposed to the mass media and the fleeting symbols of urban modernity rather than to books and critical discussion. They find their entertainment in comic strips, movies, and television, in star cults, disco-dancing, fashions, fads and fancies.

Consumer culture and environment

As new, upwardly mobile people, this new generation carry little cultural baggage, and they adapt themselves easily and uncritically to the new urban environment. The time when they came recognizably to the fore – say, the early 1970s – coincided with the advent of the high-rise buildings and the institutionalization of mass consumer culture. When the skyline went up, the street level was invaded by Colonel Saunders and Ronald

McDonald, or the local equivalents, such as Jollibee and Big Mac. All types of exotic fast food, from pizza to sushi, vie with their local counterparts, from *bibingka* rice cakes to *lugaw* rice gruel. Simultaneously, tape technology and the proliferation of commercial broadcasting exposed the public to the standard musical fare fashionable at the moment, while the air-conditioned shopping malls began to put their mark on leisure time patterns. Ready-to-wear clothing was replacing handmade garments and began to standardise urban dress, while advertising, especially on television, but also in lifestyle and fashion magazines, directed the taste of the public. At the time when the new middle class emerged, a commercial mass culture established itself, geared to this new stratum's – diverse – cultural needs: blue jeans and a nondescript watch for the less lucky, brand-named shirts, a Seiko and a car for the more successful.

The metropolitan environment appears to be moved by money, business, economic development, and technocratic 'efficiency', yet the goods that these entities produce are divided in a highly unequal way. While many experience the discomforts of public transport, others spend their commuting time in their private vehicles. But it is fair to observe that everyone shares in the shortcomings of organized public life, such as the stalling traffic, the power-cuts, the defective telephone service, the water shortage, erratic law enforcement, and so on. All these public problems provide good opportunities for politicians to take up issues and to promise better conditions, but trust in these so-called traditional politicians, all in pursuit of their private or dynastic interests, is notoriously low: the political process is breeding cynicism, or indifference at best. The public sphere appears to be uncontrollable, beyond the power of the state; it is an open space, a domain in which everybody fights his own individual battles and tries to survive. With money as its prominent measure, it is an amoral sphere in which competition, naked force, venality, poverty and extreme wealth, crime, militarization, *coups d'état*, power struggles, and violent media entertainment are most comfortably ensconced.

School and press

Interestingly, school and press have little to contribute to ameliorate the negative perception of the public world. Social studies in schools are presented in an artless way, devoid of theory, vision, system or historical perspective. Education seemingly does not offer the conceptual means to come to grips with history, society and government. It offers incoherent facts and information whilst constantly depicting the American occupation as the golden age of the nation's history. As soon as the Filipinos themselves take over, degeneration sets in while disorder, corruption, fraudulence and violence become normal. This is illustrated by the refusal on the part of the political elite to come to terms with the historical past. This allows individual presidents to betray expectations raised by the anti-colonial sentiments. In brief, the public sphere is in shambles (Mulder 1994b: 475–508).

The civics component of social studies, which is especially emphasized in elementary school, is not very helpful either. The public sphere is portrayed as an area in moral decay that should be redeemed by the Constitution of 1987 and, more particularly, by the righteous conduct of the populace. If everybody knows and fulfils their duty, society cannot be anything but an ideal place. This ideology places responsibility for the common good conveniently on individual shoulders, which is not very helpful in bringing transparency to the public domain of wider society.

The pervasive sense of malaise and powerlessness with which the students are confronted is later reinforced by a relentless stream of newspaper negativism composed by editors and all sorts of columnists, who never cease to comment on the destructive and confrontational political process, the absence of a notion of the public interest, the weakness of nationalism, the rape of the environment, and the miserable state of the economy. Apart from such man-made disasters, nature and the Earth itself are generous in contributing their share of misery, compounding the bleak picture of the wider world. Be this as it may, if the press functions as 'the mirror of the public world', its negativism and 'Philippines-bashing' do appear as part of its culture (Mulder 1996).

Political legitimation

The evolution of the public sphere and its legitimacy

The current negative image of the public sphere may be seen as one of the factors conditioning the culture of the new middle class, but such a narrow focus on the present would avoid drawing attention to the prevailing ambiguous perception of this sphere that appears as a general problem of the history to which all Filipinos are heirs. This warrants a short excursion into the past to see how such a perception came into existence and has been constituted. This kind of reflection also allows a valuable precision to the notion of public sphere.

A public sphere contrasts with the private sphere of family and community life yet, whereas the latter is universally given, the first is not. In less differentiated societies, the community is surrounded by uncharted territory and nature into which forays are made, but these have little to do with the idea of a public sphere. To put it in Habermas's parlance, a public sphere consists of the two independent subsystems of state and economy in which the life world is embedded. It only comes into existence when the spheres of state and economy have differentiated from a simpler form of social organization. Apart from its own logics and expediencies, such a public sphere may generate its own culture, as expressed in the public opinion that is fed by free media and other institutions of civil society. To put it differently, the culture of the public sphere is the overarching world of ideas concerning society-in-the-abstract, that is, Gesellschaft.

Historical knowledge of *baranggáy* society provides a vivid illustration of the above considerations: as relatively small communities they were organized around a chieftain (the *datu*). The strength of personal loyalties to this *datu* spelled the strength of the community. Hierarchically organized, people felt that they depended on the patronage of their leader. The functions of production, administration, religion, and so forth, were an integral part of a way of life to which analytical abstractions, such as economy, state, and sphere of the sacred, did not apply. In simple terms, life in such communities was organized along familial lines, people experiencing their social life directly, personally. Beyond this particularistic sphere of life one found other such communities which may have been seen as friends or strangers, trading partners or enemies, but no overarching public sphere brought such communities together. In other words, no higher, organized polity transcended the individual communities. The image of the outer world became a field of opportunity, to be appropriated as need be; a field where the individualistic adventurer finds his gain. He is not responsible for that wide area; it is not organized; it is nobody's and everybody's territory; it is open space devoid of a public, of people belonging together in-the-abstract.

Such was the situation upon Legazpi's arrival, when people and territory were gradually brought together under Spanish colonial authority. It imposed higher, less direct layers of government whose super-patron was the King of Spain. The line of patronage was extended. Yet, because this encompassing sphere was colonial and imposed, also autonomous and beyond volition, exploitative and authoritarian, it was low in legitimacy and difficult to connect with local circumstances. For it to be accepted, two things were necessary: the cooperation of the former ruling families, and missionary activity.

By the extension of privilege, the Spaniards were successful in coopting the *datus* as their henchmen. As the brokers between the alien sphere of colonial government and the local populace, their position changed. They became less dependent on local acceptance and derived a measure of support from the new overlord. Somehow they moved into a new space. While their traditional authority was based on effective leadership that legitimized their position, from then on their position became dependent on their capacity to gather taxes effectively and to maintain peaceful conditions which would not interrupt colonial rule.

There were those who fled from colonial impositions, and we may assume that the population who remained only grudgingly accepted their new situation. The new dispensation was probably more exploitative than the former, and their only role in it was to obey. The ones who ultimately profited were the native petty ruling class – the *principalía* – who operated in a wider world of government and who were allowed to appropriate a part of it. Seen from below, the expanded sphere of government belonged to the petty rulers who could extend their private space and profit from their privilege; their governance had little to do with care for the common good. This is not to say that the native *principales* were necessarily indifferent to the welfare of the people under their sway, but to maintain their position they had to develop the skills of balancing the demands of the powerful overlord with their own interests, and, in the last place, with those of the population. Their position, and their interests, plausibly induced a political culture of artfulness and deceit that aimed at safeguarding their own interests most of all (Corpuz 1989: xii–xiii).

Colonial legitimacy was best served by the missionary friars who, in accord with official policy, herded the population 'under the bells'. While this incipient *pueblo* (township) society was dominated by the parish priest, he needed the alliance with the *principales*, who would grow to become its elite. Be that as it may, over time Christianization refocused the fulfilment of religious needs on priest and church, which lent an aura of accepted authority to the religious establishment. This is not to say that the native understanding of its messages followed the Spanish pattern; far from it, and it is of interest to stress the counter process of Filipinization of Spanish Catholicism (Mulder 1992b).

While the imposition of colonial rule placed government, and the abstract institution of the state, at a great remove from the people at large, religion remained an integral part of the life world, however instrumental it may have been for Spanish dominance. It is only much later that the sphere of the economy began to separate from direct experience. At the beginning of this differentiation stand the economic changes initiated in the second part of the eighteenth century and continuing well into the nineteenth century, when international demand stimulated agricultural development, and thus the quest for landed property, and trade. While these developments were partly presided over by Spanish colonial monopolies, they also stimulated the rise of the recognizable ancestors of the contemporary oligarchy. When, subsequent to the British occupation of Manila, the Chinese were expelled, and most immigration ceased, the remaining Christianized Chinese-Filipino *mestizos* stayed on, and their economic success fused with the economic and administrative privileges of the *principales*. As a result, a new entrepreneurial class came into being that appropriated economic opportunity, many of its members growing rich, and sometimes very rich. It should be noted that they were driven by the quest for money; they were economic men, and politically virtually powerless. In terms of government, the colony remained the preserve of the Spaniards, and their perceived interests defined the public interest.

Gradually, a clearly felt opposition grew into existence between the ruling class from the Iberian Peninsula and those who identified with their native islands. This prepared the ground for the modern nationalism which arose in the nineteenth century. While it is most improbable that this coincided with Spanish intentions, in retrospect it is safe to conclude that the colonial government did all it could to promote its growth. The exacerbation of the secularization issue from the time when the Spanish empire had lost most of its American possessions, came to a head with the judicial murder and martyrdom of the Philippine-born priests Gomez, Burgos and Zamora in 1872 – a shocking event that galvanized the Filipino national sentiment into being.[1]

Born of colonialism, the Philippines was the first Asian nation to rebel against the mother country and to fight successfully for its freedom. To do so required more than an abusive regime; it also demanded the unifying idea of nationalism. This early Philippine nationalism discovered the Filipinos as a people and strove for independence. Accordingly, the elite leadership of the Revolution considered themselves as born rulers while shaping independence in their own best interest; the others were to follow. The non-elite Katipunan did not envisage a social revolution; it was the fight against Spain that mattered, not the shape of tomorrow; it was the realization of the mystique of nationhood, of brotherhood; it was nationalism as religion.

The importance of all this in relation to the public sphere is that alien, colonial transcendence had made way for a native expression of belonging together, of being members of a nation, of having a shared territory and history, of having a particular identity. The rapid succession of events around the turn of the century prevented the Filipinos, however, from substantiating their nationalism in their own public institutions, such as government, planning and the projection of a course into the future, an educational system, and so forth. The First Republic could not institutionalize itself; what had grown into being was a desire for independence and a spirit of nationalism that centred on the cult of the motherland rather than on the politics of nation-building. It was in this milieu that the Americans imposed their colonial dispensation.

The Americans had many ideas about the construction of a public sphere while imposing a 'colonial democracy' that allowed the elite to become an effective ruling class; the gradual transfer of political power to them brought them so much economic gain that 'nationalism' and 'independence' faded as priorities. After all, they had come into existence as economic men, and had learned to see the country as a private preserve and field of opportunity. In this sense, there is a lot of cultural continuity between *baranggáy* conceptualization and the present, and in economic relations between the nineteenth and the twentieth century. This continuity also affected the economically powerless. Their exploitation remained a mainstay of production, and the introduction of socialist and Communist thought did not achieve citizenship for them. They had no stake, or influence, in the public sphere that was appropriated by an increasingly self-confident and America-oriented oligarchic elite. The best option for the oppressed farmers and labourers was resistance, unrest, strikes, and messianistic or political uprisings; they remained marginal to the system.

The continuity in economic life was accompanied by the evolution of novel political institutions that were soon shaped in the image of previous practice. The introduction of the 'pork barrel system' and its 'spoils' remodelled politicians into patrons who were less concerned with good government, and more with the distribution of benefits, the building of dependency relationships and political machines. This emphasis on patronage was in glorious continuity with the past, and fitted in well with the economic interests which began to dominate public life. The language of this public sphere became English, as such underlining its appropriation by an educated elite far removed from the ordinary people who expressed themselves in the vernaculars.

Whether the substitution of Spanish by English as the language of the public sphere is symbolic for the cultural discontinuity in social life may be debated. Other factors might be more obvious. The long Spanish period resulted, at least among the educated elite, in a historically created consciousness of identity and character that definitely played a role in the articulation of nationhood. Catholicism, a European intellectual orientation, and a vast quantity of other Spanish cultural goods were an un-questioned part of their heritage. What the Americans succeeded in doing was more than substituting one language for the other. By relegating the islands' Spanish history to insignificance, by portraying it as a period of protocivilization, and by aggressively promoting the idea of the superiority of American culture and history, the United States colonialism was not only successful in brainwashing its new subjects, while making them eager to associate themselves with progress and modernity, but it also destroyed the sense of historical continuity and the identity that comes with it. In terms of nation-building, the gravest consequence was the destruction of history and its replacement with a forward-looking orientation. This cutting loose from the social and cultural past led to the glorification of the American period, as can be observed in the schoolbooks of today and the weakness of the nationalistic sentiments among the people at large.

With the country's own history being denied, and in spite of Quezon's battle cry for 'Immediate, complete, and absolute independence', America's colonial tutelage led to a vacuous nationalism and a primary political conception of the public sphere. The Filipinos whom the American administration coopted were the pliable descendants of the elite and, once given power, they went all out for the playing of politics per se, divorcing its practice from the democratic demand for good goverment (Paredes 1988). Quezon's statement that he preferred 'a government run like hell by Filipinos to one run like heaven by the Americans' turned out to be prophetic, since that was precisely what the country got.

The founding fathers of the Commonwealth and the Third Republic cannot be credited with attempts to fill the public sphere with nation-building or social reconstruction. Elitist by nature, they cared chiefly for their own interests and had nothing to gain from popular mobilization. In their mercenary culture of the public sphere, the localization of the American institutions of government became a travesty of democracy. To keep the people in their place, they subscribed to the idea that the good order of society follows from the moral conduct of individuals who are aware of their obligations, as spelled out in Quezon's Code of Citizenship and Ethics.[2]

From this brief excursion into colonial history, we may safely conclude that state and economy were instituted and dominated by colonial powers and their elite political successors. Because all of these were primarily self-serving, their legitimacy was low and at the same time the public sphere was denied its chance to evolve into a moral and consensual entity, perceived as a nation. The immediate politicization of the civil service, meaning the rewarding of political supporters with bureaucratic office, from the time Quezon took power (McCoy 1988: 144–68), further destroyed whatever legitimacy the state and its instruments could have inspired. What took over, and even more visibly so after the destruction and demoralization of the Pacific War, was an undisguised jockeying for, and exploitation of, power.

Colonial mentality

The granting of independence to the Philippines in 1946 had little to do with the war and its aftermath; it was already underway at the time the Americans were hailed as liberators of the Japanese occupation. Rather than the culmination of a fight for freedom, it inaugurated a period of profound dependence on the United States, culturally, economically, and even politically. In terms of a culture of the public sphere, it signalled a malaise from which the country still has to recover. In the late 1950s, Claro M. Recto reminded his countrymen that continuing dependency on Washington and the Pentagon was a betrayal of their nationhood, and that this begging attitude was an insult to independence. In his newspaper columns, Soliongco also tried to contribute to awareness and spiritual independence, while it was Constantino who got to the root of the problem with his famous comments on the (neo)colonial miseducation of the Filipino.[3] While these critical commentators stimulated the subsequent discussions about social, cultural, and nationalist reconstruction, their criticism fell largely on deaf ears because, in spite of the havoc wrought by war, rebellion, and confrontational 'traditional' politics, they were up against a highly self-satisfied smugness that considered the Philippines as the most advanced of Asian nations.

Into the 1960s, the Philippine economy appeared as one of the most robust in Asia, with an advanced, educated entrepreneurial class and privileged access to the American market. People still believed in this special relationship with their senior partner across the Pacific. Culturally, they saw themselves as part of Western civilization; the third largest English-speaking country in the world; the only Christian nation in Asia; the showcase of democracy in a region ruled by strongmen; the bridge between East and West. This Western orientation, the emulation of the American model, and their unquestioned position in Asia by its own logic became the source of bedevilling dilemmas that, at the time, were seldom recognized. An early protestation can be found with Corpuz who objected to the imposition of American ethical measures to judge Philippine conditions; Filipinos should realize and develop their own standards.[4] But then, where to begin? How to create consciousness of nationhood, of history, of identity? In its aimless quest for progress and future, and glorification of the American period, society at large had drifted far from any historical moral moorings.

The sham of it all, the perfidity of politics, the depth of social cleavages, and the question of identity all burst into the open during Marcos's expedient exercise of power. In 1969, the New People's Army (NPA) was established; the disillusionment caused by Marcos's fraudulence and mendacity took violent shape during the protracted protests of 1970) which became known as the First Quarter Storm. In view of the president's personality, it logically led to the declaration of martial law in 1972. One by one the 'achievements of civilization' were destroyed; civilian control of the army gave way to the nightmare of militarization; dictatorship replaced democracy; censorship killed the free press; propaganda took the place of information; the 'independent' judiciary became a travesty of justice; the superficiality of the Christian tradition was exposed; the introduction of bilingualism in school eroded the understanding of both English and Filipino (Tagalog); education was made the instrument of propaganda and an ahistorical process, resulting in that amazingly meek and blank generation of martial law babies; economically, the country became the sick man of Asia; and more than ever violence throve.

The short-lived euphoria of the People Power Revolution during the uprising against the Marcos regime, its pride in nationhood, and desire for national reconciliation were soon smothered in massacre, militarization, vigilantism, the unsheathing of the sword of (civil) war, violent *coups d'état*, economic stagnation, and the restoration of traditional politics. Adding to the sense of abandonment was a constant stream of natural and man-made disasters, from 'the greatest civilian maritime mishap ever' to devastating earthquakes and volcanic eruptions. The country appeared to be God-forsaken and filled with more problems than man could ever hope to solve, so reinforcing both a 'colonial mentality' that believes that the world 'out there' is superior, and thus worth emigrating to, and the perennial orientation to the private, moral sphere of family and small group.

Identification

The culture of the public sphere of the new middle class

It is difficult to identify with a public sphere that is seen as essentially fiendish and unamenable to good intentions, let alone to participate in it constructively and responsibly. For a long time it has been the area of traditional politics, of the tricks and deceit of colonial rulers or the neo-colonial oligarchy. Regarding the country as their private preserve, these people had no interest in creating a vibrant public sphere of participating citizens. While some blame this on the absence of nationalism among the political elite (Corpuz 1989, Vol. II: 534–83), it should be seen as a perennial feature of the Republic, whose leaders, unless charismatic, have always been low in popular legitimacy, more concerned with personal power and 'party' interests than with the democratic demand for good government. Basically, the public sphere is ruled by political and economic considerations, and not by the idea of the public interest. For most people, therefore, it is a sphere to defend oneself against, or to profit from; one's real life and identity belong elsewhere.

All this fits the picture of the contemporary mass society. People do not really participate in urban affairs; they are simply there, much as one is in a forest without participating in nature. The new urban middle class have no tradition of being involved in 'public' affairs. Before, they were taken care of by known leaders, such as patrons or political bosses; at present that still seems to be the case, although less clearly so. The current picture of society is vaguer, more complex, less personal, less tangible, less easy to comprehend. The place seems to be run by the political and business establishments that dominate a relatively soft state, and if it is in a mess, venality, rapacious politicians, moral decay, American interference, Chinese businessmen, or twenty years of Marcos must be blamed. Even so, on occasion the new and old middle class joined in protest, such as in the period between the assassination of Ninoy Aquino (21 August 1983) and the demonstrative events of February 1986 that propelled his widow to the presidency. Yet, these protests were without a programme, and how to bring about a meaningful social transformation is beyond the imagination of the members of the new middle class.

The academics, progressive intellectuals and clerics who are still preoccupied with programmatic social change are the heirs to the debates of the 1960s, when social reconstructability held promise. Organized in cause-oriented associations and non-government organizations (NGOs), they clamour for land reform, social justice, human rights, empowerment of the weak, emancipation of women, educational reform, demilitarization, and other issues (Mulder 1992c: 165–9). On occasion, they are quite vociferous, yet they fail to trigger a broad societal discussion and, especially since the end of the Marcos era, they have become peripheral; the impact they make on the political process is negligible. Even if the points they raise were to be taken seriously by the political centre, it is doubtful whether politicians have the political will or whether the administration has the executive power to bring about meaningful reforms.

Individual-centredness

The ungovernable nature of the public sphere reduces policy declarations to mere rhetoric and stimulates the wisdom of a dogged individualism with everybody caring for himself and his dependents (*kanya-kanya*), a tendency that is reinforced by the appeal to spend money, to consume, and to buy the status symbols that mark one's individuality. Where society is lost sight of, its component individuals come to the fore, and if the experience of metropolitan mass society is unsatisfactory, the wholeness of the personal group with its primordial identity functions becomes extraordinarily important. In this conception, private life centres on solidary groups and public life on outstanding, or anonymous, single individuals.

Society that matters is society that is experienced directly, tangibly. It is not about abstract issues, institutions and structures, but about people, concretely known (Mulder 1994c: 80–90). Social life is about gender, status, honour, protection, reciprocity and obligation, gratitude and patronage, and, above all, about being related and known to each other. All these contribute to the direct experience from which the moral obligations and identity arise that make social life stick together. It is in this society that every person's experience is real; it contrasts with the unknown, impersonal and intangible sphere of law, or the idea of the public-in-the-abstract, the 'generalized other'.

The new urban society surrounding private life is wide open to the world; it is a part of a postnational, global society of metropolitan centres. It is not subject to any ideology or ethical system other than the rules of political expediency and business. Because of people's dependence on it for survival, it intrudes into private life, and causes the frustration that sometimes results in protests against the wheeling and dealing going on in the worlds of politics and corporate decision-making. The objections arising from the new middle class are typically phrased in ethical rather than ideological terms. If everybody leads a moral life, the public world of wider society should be in good order. This belief in the force of right-eousness is held to be illustrated by the idea that is was moral protest that forced Marcos out. But apart from that great event, such ethical postures are not expressed in demonstrative gatherings but in newspaper columns and letters to the editor, in values education courses, in sermons and ex-hortatory speeches, emphasizing decency of manners, sacrifice, and personal virtue.[5]

Contemporary urban society with its thriving service sector certainly produces job satisfaction and identity for professionals, technocrats, politicians, and businessmen. Most people, however, remain hidden and anonymous in an unruly, often violent, and corrupt, competitive environment dominated by capital. This experience of an unsafe fragile society that escapes from grasp and control drives people to find their identity in particularistic groups and associations. First of all in their families, but also in brotherhoods and *barkada* (friendship groups), sects and religious societies, in ethnically defined support groups, in civic clubs, in NGO activities, nationalistic coteries, and a variety of left-of-centre political cliques. But what appears as a highly developed civil society is not so much pressing its demands on the state and political society as turning its back on it; inwardly directed, it is expressive of personal righteousness and moral rectitude more than of transcending ideas about the common good, nationalism, nation-building, or political ideology. This is not to say that such ideas were never articulated. Communist and socialist ideas about social rearrangement have been produced throughout this century, and could mobilize groups of people from the 1920s up to the present, but they have remained entirely marginal to the political process and failed to penetrate the social imagination of the new middle class.

In the thinking of the latter, the roots of societal disorder are located in individual ignorance of correct conduct and subsequent wayward behaviour, and so, if public life is to be ordered and improved, individuals should be taught and disciplined. As a result, they propose values education to achieve moral recovery.[6] It is questionable whether there is anything to recover, whether there ever existed a public sphere, whether there ever was a Filipino citizen. Where would he hail from? The American 'preparation for self-government' with its tutelary 'colonial democracy' certainly did not result in enlightened citizenship but merely reinforced a small political elite to whom power was delegated and who, in the process, were able to vastly increase their hold on the political economy. By undermining historically-grown identity, nationalism, and participatory democracy from the outset, the Americans deliberately contributed to the creation of a hollow public sphere, an area of inter-personal rivalries, political rhetoric, artificial English, and rootless institutions of government.

The people who propose moralistic – as opposed to structural –

measures to improve public life are very much in line with their roots in family and community. In the absence of a native moral model of wider society, they can only imagine melioration of the public spheres of state and economy by applying values befitting the small group to Gesellschaft, to society-in-the-abstract. Perhaps this vision is timely indeed. The new public sphere, dominated by political and economic expediency, is morally vague, hard to imagine and difficult to identify with. What is imaginable is what is known. Therefore people rather find their security and identity in particularistic moral choices, leaving small wonder as to why there are such strong tendencies to religious revival in the Philippines, and in Southeast Asia in general (Evers 1993).

The culture of the public sphere

The primary purpose of this chapter has been to draw attention to the newness of the present urban middle class, the bulk of whose members are without a tradition of critical thinking about, and participation in, the public sphere. This class has expanded in pace with the very rapid urbanization and concomitant rapid changes in the economic opportunity structure of the past thirty years. They differ qualitatively from the more exclusive middle group that studied when going to university was still a privilege, and whose members indulged in the luxury of thinking about democratization and social reconstruction. Most members of the new urban middle class are upwardly mobile, mass-educated, and directed more to professional advancement than to ideological reflection. Besides, they acquired their skills during a period, beginning with the 1970s, of systematic propagation of lifestyles and consumption. They grew up with television rather than with books.

The culture of the new urban middle class consists of cynicism and indifference regarding the public sphere. They take it for granted rather than supporting it, and so they do not bestow a great measure of legitimacy to the state; this attitude shows an impressive continuity with the history of the public sphere and the failure to develop a culture of nationalism and participatory democracy. Although the members of the new urban middle class would like to see 'law-and-order' prevail, they tend to be socially inattentive while suffering from the urban disorder they experience without great ideas of how to improve it. This drives them to a stronger identification with family, friends, and other particularistic forms of association that emphasize individual worth while seeing personal ethical conduct as the well-spring of good society.

Other particular individual-centred choices are made in the drive for self-improvement and careerism, and in the lifestyle phenomenon that, at the centre of consumer culture, enables people to accumulate the status symbols they need to make their mark on themselves and society. These self-centred orientations lead away from ideological or theoretical attempts to come to grips with their public environment, which as a result remains vague. Wider society is an area to watch on television and read about in the newspaper, which does not stimulate in itself the development of a civil culture of the public sphere among the members of the new urban middle class. The conclusion is that only minimum demands appear to be evolving from this stratum on the public sphere of state and economy.

1.   The repression of the demands for reform, culminating in the terror of 1872, united the people of the colony against the Spanish-born *peninsulares*. While the word Filipino was used formerly for a Spaniard born in the islands, it then came to cover all people native there. The three murdered priests, popularly abbreviated as Gomburza, symbolize this: Mariano Gomez, who devoted his life to keep his parish out of the hands of the peninsular friars, was a native of the islands, with possibly a tinge of Chinese ancestry; Jacinto Zamora was a Spanish *mestizo*; Jose Burgos was a Spaniard born in the Philippines. All were secular clergymen. See Corpuz (1989, Vol. II: 1–38, esp. 29–31).

2.   This Code urges people 1. to have faith in the Divine; 2. to love their country; 3. to respect the constitution; 4. to pay their taxes; 5. to safeguard the purity of suffrage; 6. to love and respect their parents gratefully; 7. to value their honour; 8. to be truthful and honest; 9. to lead a clean and frugal life; 10. to live up to the noble traditions of their people; 11. to be industrious; 12. to rely on their own efforts; 13. to do their work well; 14. to contribute to the welfare of their community; 15. to use Philippine-made products; 16. to use and develop the country's national resources while not trafficking with their citizenship.

3.   About Recto, see Constantino (1971); about Soliongco, Constantino (1981); about (neo)colonial education, Constantino (1966).

4.   Already in 1960, Corpuz argued that in Filipino politics nepotism is ethically normal and that party loyalty is subject to family-based interests. That is why 'We do [should, NM] not judge ourselves by the irrelevant idiosyncracies, eccentricities, and even wishes, of alien nations.' Corpuz (1969: 6–18).

5.   For a number of examples, see Mulder (1994d).

6.   Creating morally-aware individuals as the mainstay of a good society is the purpose of the compulsory study of the Constitution in school; it emphasizes the noble intentions of the Filipinos as a people, plus the rights and duties of the citizen. Next to this, values education is an integral, and important, part of the social studies curriculum. Aimed at the populace in general is the moral recovery programme proposed by Senator Leticia Ramos-Shahani in 1988; it was officially inaugurated by her brother, the incumbent president, in 1993. See Instructional Materials Corporation (1988).

Bibliography

Caoili, Manuel A. (1988) *The Origins of Metropolitan Manila. A Political and Social Analysis*, New Day Publisher, Quezon City

Constantino, Renato (1966) 'The Miseducation of the Filipino', in Renato Constantino, *The Filipinos and the Philippines and Other Essays*, Malaya Books, Quezon City

Constantino, Renato (1971) *The Making of a Filipino*, Malaya Books, Quezon City

Constantino, Renato (ed.) (1981) *Soliongeo Today*, Foundation for Nationalist Studies, Quezon City

Corpuz, O. D. (1969) 'The Cultural Foundations of Filipino Politics', in

J. V. Abueva and R. P. Guzman (eds) *Foundations and Dynamics of Filipino Government and Politics*, Bookmark, Manila

Corpuz, O. D. (1989) *The Roots of the Filipino Nation I, II*, Aklahi Foundation, Quezon City

Evers, Hans-Dieter (1993) 'Religious Revivalism in Southeast Asia', Special focus issue of *Sojourn*, **8**(1)

Instructional Materials Corporation (1988) *Building a People, Building a Nation*, A Moral Recovery Program, Quezon City

McCoy, Alfred W (1988) '*Quezon's Commonwealth: The Emergence of Philippine Authoritarianism*', in Ruby R. Paredes (ed.) *Philippine Colonial Democracy*, Yale University Southeast Asia Studies, Monograph Series, 32, New Haven, Conn.

Mulder, Niels (1992a) *Individual and Society in Java. A Cultural Analysis*, Gadjah Mada University Press, Yogyakarta

Mulder, Niels (1992b) 'Localization and Philippine Catholicism', *Philippine Studies*, **40**(2)

Mulder, Niels (1992c) *Inside Southeast Asia: Thai, Javanese and Filipino Interpretaions of Everyday Life*, Editions Duang Kamol, Bangkok

Mulder, Niels (1994a) *Inside Thai society: An Interpretation of Everyday Life*, Editions Duang Kamol, Bangkok

Mulder, Niels (1994b) 'The Image of History and Society (in Philippine high school texts)', *Philippine Studies*, **42**(4)

Mulder, Niels (1994c) 'Filipino Culture and Social Analysis', *Philippine Studies*, **42**(1)

Mulder, Niels (1994d) *Philippine Public Space and Public Sphere*, Working Paper No. 210, Southeast Asia Programme, Sociology of Development Research Centre, University of Bielefeld

Mulder, Niels (1996) *Everyday Life in the Philippines: A Southeast Asian Interpretation of Filipino Culture*, New Day Publishers, Quezon City

Mulder, Niels (1996) ' "This God-forsaken Country": Filipino Images of the Nation', in Hans Antlöv and Stein Tonneson (eds) *Asian Forms of the Nation*, Curzon Press, London

Parades, Ruby R. (ed) (1988) *Philippine Colonial Democracy*, Yale University Southeast Asia Studies, Monograph Series 32, New Haven, Connecticut

CHAPTER 6

Ethnicity, class and human resource management in Singapore

David Drakakis-Smith

Hettne describes ethnicity as a fluid and elusive concept involving not only a wide range of objective factors, such as race, language, culture and religion, but also subjective considerations that comprise a 'felt' combination of objective factors which are fluid over space and time (1993: 123–49). Brown also identifies two dimensions to ethnicity within the context of development. First, it can be seen as primordial and cultural, making people 'naturally' ethnocentric; second, it can be considered as a less conscious membership of a cultural group, the ethnicity of which becomes focused only in response to external threats – in short an 'interest group' (1994). In most cases, the real world is usually a constantly changing combination of these two dimensions (Burja 1992: 347–61).

In the precolonial world ethnicity was the principal foundation on which political entities were created (Lemon 1993) but colonialism paid as little heed to such patterns as it did to physical geography when recasting the political map in the nineteenth and twentieth centuries. In this process ethnicity and race continued to be fluid concepts in which groups were categorized by deliberately vague terminologies into 'Asian' (in East Africa) or 'Chinese' (in Southeast Asia) when in reality such groups contained quite different and often antagonistic language or cultural groups (Wallerstein 1989: 1–18; Lubeck 1992).

It is somewhat surprising, in view of this manipulated dimension to colonialism, that postcolonial theories of development have neglected ethnicity, despite its central role in countries such as Lebanon, Sri Lanka or Yugoslavia. Dudley Seers' (1969) much admired definition of a human-needs centred development made no reference to ethnicity and more recent attempts to update his list of essential items have incorporated gender and environmental issues (Thomas and Potter 1992: 116–41) but not ethnicity. Perhaps it is Hettne's 'elusiveness' that perturbs the economists who have dominated the formulation of development strategies in the second half of this century. But if we accept Brown's definition of ethnicity as one of several forms of association through which individuals pursue interests relating to economic and political advantage, then it is possible to insert ethnicity into mainstream development (Brown 1994:

xii). This is particularly the case if, as Hettne suggests, ethnicity is seen as part of what he terms the 'nation-state project' – the creation of a national entity which is at the core of the development process. He identifies the common objectives of nation building as:

- exclusive political/military control over a certain territory
- defence of this territory against external claims
- creation of internal material welfare
- creation of internal political legitimacy
- creation of substantial internal cultural homogeneity (Hettne 1993: 6)

This implies that in plural societies state power, legitimacy and modernization often become associated with a dominant ethnic group. If the process slips too far in this direction, the objective may become the creation of an ethnocratic state, as in Malaysia or Sri Lanka. Of course, this is not a situation which has been confined to the postindependence period, as far as developing countries are concerned. Indeed, the creation and consolidation of the colonial states of the nineteenth and, particularly, the twentieth centuries were clearly part of the same process. Nevertheless, tensions released by independence and exacerbated by uneven development have accelerated conflicts between ethnic groups within the development process. Moreover, during the last twenty years or so, increasing mobility, forced or voluntary, has become characteristic of both the national and international scene, further complicating what were often already complex ethnic situations.

Southeast Asia has experienced enormous population movements across the years, both prior to and as a consequence of colonialism, including what Hodder has termed the Chinese diaspora (Hodder 1992; Winchester 1991). At independence colonial territorial boundaries that had had little relevance to physical or ethnic geographies became confirmed as the political stage on which Gunnar Myrdal's Asian drama was to be played out, and on which major and minor players struggled for pre-eminence. Brown has observed that even within the relatively small region of Southeast Asia, there are substantial variations in ethnic politics. In some countries, such as Myanmar and Malaysia, ethnocentrism dominates. In Myanmar, however, this is complicated by an ethnic territoriality which sees the dominant Burmans constantly challenged by regional secessionist movements. Thailand too has an ethnic dimension to its development but in its case the exploited minorities on the periphery are not dominated by a single ethnic group at the centre, rather by an often uneasy Thai–Chinese alliance. Some states, such as Malaysia, delayed the elevation of ethnicity to a major role in the national development process (the nation-state project). Here, under the banner of redressing ethnic inequality, one ethnic group, the Malays, has been favoured over others, the Chinese and Indians, in political, economic and cultural change since 1970 (Dwyer and Eyre 1993; Williams 1993). As noted earlier, these ethnic labels themselves conceal, perhaps deliberately so, quite considerable sub-ethnic differences and even antagonisms (Lubeck 1992), and have frequently been employed to obscure equally problematic class cleavages (Abrahams 1985; Ghaffer 1987). This is an important issue which will be discussed later.

Singapore, the near neighbour of Malaysia, with the same basic ethnic ingredients, seems to have chosen to structure national development in quite a different way; one more analogous to Hettne's cultural pluralism set within 'another development'. Cultural pluralism encompasses mutual respect between ethnic groups, the right to speak separate languages, practise different religions and the like. Consequently, according to Hettne, it challenges the nation-state project because cultural and political boundaries rarely coincide. Broad-based intra- and inter-national tolerance may undermine the pre-eminence of the dominant group. Teo and Ooi contest Hettne's view in the context of Singapore, suggesting that the republic's government has sought ostensibly to maintain, but in effect to supplant, ethnic values by a national ideal structured around economic growth. Ethnic or cultural plurality has been permitted, according to Hettne's definitions, but the Singapore government has sought to create a synthetic amalgamation of cultures in which the more traditionally-based ethnic ones are subordinate to a national ethos based on economic rationality and a meritocracy; a combination which Hettne suggests cannot work, but which apparently does in Singapore. Indeed, according to Teo and Ooi, the Singapore government has 'strictly enforced racial harmony' which seems to be a contradiction in itself. True harmony cannot be enforced (Teo and Ooi 1993: 2).

Brown has attempted to explain Singapore's ethnic policies with reference to the concept of the corporatist state. Despite the connotations of this term, the corporatist state is essentially a reconstructed feudal system of government in which an autonomous elite seeks to organize diverse interest groups in terms of their relationships with one another and with the state itself. This elite does not comprise a bourgeoisie promoting its own class interests, but is a small group of legal, managerial and economic experts acting in the interests of the state per se. In developing societies, Brown argues, the corporatist state has emerged from the need to curtail populist participation in the political process when substantial sacrifices are demanded of the proto-proletariat. This 'delayed gratification' (Hsaio 1992: 17–32), demands active cooperation from the masses and is based on three principles: first, absolute loyalty to the nation-state with subgroup loyalties considered as undermining this; second, the creation of a tiered cultural identity comprising the various values of the main interest groups, which are seen as different but compatible; and third, the interests of the groups which are recognized as legitimate are licensed and function as a means of political control as well as an avenue for cultural expression.

Most discussions of the corporatist state are structured around economic interest groups but Brown extends the concept to the incorporation of ethnicity in the political economy of Singapore because of the association between ethnicity and labour roles in the colonial period, and because of the geographical factors associated with Chinese immigrant communities in a Malay region. Ethnicity is thus seen as generating subnational loyalties that challenge the nation-state, as Hettne notes. However, in Singapore the state has chosen to use these cultural identities as an important component in the creation of a national identity or ideology – in Singapore's case this is an 'Asian' rather than a 'Western' industrial society. Finally, ethnic groups are used to promote a legitimized but restricted form of political participation, one which gives the satisfaction of an input into the political process but which also permits the state to control the constituent activities. Brown argues that throughout the 1980s and 1990s these corporationist tendencies have increasingly dominated political development in Singapore and have been seen as an essential adjunct to the evolution of the industrial-export economy.

Not all commentators agree with this interpretation, with some support existing for the view that Singapore's economic development has produced a middle class which has become economically and politically independent and will thus generate liberalization (Rodan 1989). Ho has, however, argued that the middle class in Singapore is obsessively preoccupied with *Kiasu*, or selfish, competitive materialism, and that it has been successfully incorporated into the corporatist system and made politically neutral (1989: 670–98). This is a perceptive observation which will be raised again later in discussion, although it must be noted that *Kiasu* is more associated with the Chinese ethnic group, the middle class of which tends to self-identify with regard to wealth rather than a set of values (Chiew 1991: 138–82).

The purpose of this contribution is to examine the bundle of apparent contradictions that is evident in Singapore and to assess not only the extent to which the republic has been able to create a plural society but whether the constituent ethnic groups are playing the roles expected of them in the nation-state project. In Furnivall's plural society there is no common will (Furnivall 1980: 86–96); in Hettne's approach this is not addressed; but in Singapore's it is central. The current Prime Minister has on many occasions echoed his predecessor, Lee Kuan Yew, in reminding Singaporeans to lift their horizons beyond ethnic matters and pull together for 'Team Singapore' because individual community problems 'cannot be resolved in isolation. They can only be resolved in the context of the nation' (Teo and Ooi 1993: 9). But as societies develop and modernize within capitalism, so do class cleavages, and although this chapter will be predominantly concerned with ethnicity, it cannot ignore the way in which the two have interacted.

Economic growth and human resource management in Singapore: building the nation-state

Prior to full independence in 1965 Singapore had been an archetypal colonial city functioning primarily as an entrepôt for Southeast Asia as a whole and for British Malaya in particular. As with many colonial cities there was a strong relationship between ethnicity, function and status. Europeans occupied senior positions in commerce, administration and the military; various Chinese clans dominated regional and local trade and commerce; a limited Chinese proletariat had developed; Indians were also strongly involved in small-scale trading, whilst a relatively small number of Malays were engaged in fishing, farming and domestic work.

Not only did ethnic stratification exist in economic, social and cultural terms, it had from the inception of Singapore been an integral part of the built environment. Sir Stamford Raffles had laid out the street plan for his new settlement along both occupational and ethnic lines. It was, of course, not uncommon to find contemporary cities divided into distinct ethnic enclaves; what marks Singapore out as somewhat different is the extent to which this spatial/ethnic association has been maintained over the intervening centuries. Indeed, so persistent and significant have been these place–people linkages that the bulk of the contemporary conservation programme in Singapore has an extremely strong ethnic underpinning. Kong and Yeoh provide an excellent discussion of this process, revealing that the first four major conservation projects in Singapore have focused on districts with a clear ethnic identity, i.e. Chinatown, Little India, Kampong Glam (Malay) and Emerald Hill (Peranakan).

Of course, it can be argued that conservation along ethnic lines may emphasize differences at the expense of the 'glue which could bind a multi-ethnic, multi-cultural society together' (Kong and Yeoh nd: 17). In many ways, however, making such claims obscures some of the real issues. In the first instance, it is obvious that most of the conservation projects in Singapore have produced artificial environments which fail to retain either the character or residents of their earlier role. Wealthy families, up-market businesses and tourists now dominate much of these conservation districts. Furthermore, an important part of urban conservation in Singapore is the colonial heart of the old city, now variously known as the 'Civic and Cultural District' or 'Heritage Link', which 'boasts a collection of European neo-classical landmarks', such as City Hall, the Supreme Court and Raffles Hotel (Kong and Yeoh nd: 25). It is a sign of the confidence and/or pragmatism of the present government in Singapore that the nation-state project does not feel threatened by the colonial legacy in the built environment, reinvesting it instead with an overlay of national pride. Indeed, for many Singaporeans it seems that the grandeur of the colonial built environment is more worthy of conservation than more culturally modest features, such as shophouses.

Interesting as this discussion is in relating ethnicity to contemporary development in Singapore, it has taken us too quickly through the evolution of the modern state. This began in 1965 in particularly unpromising fashion with a city the population of which was rising very rapidly (it was a prime target for Chinese economic refugees from cities throughout Southeast Asia), the crumbling built environment of which was aptly mirrored by the tacky society set within it by Paul Theroux in his novel *Saint Jack*, and the economic prospects of which were not rosy. The new government, stressing the fragility of the state, emphasized the unity within its ethnic diversity and promised that all groups would be treated fairly. This did not mean equally since the Malays were given special attention, partly because they were the least advantaged group and partly in recognition of regional 'realpolitik' (Brown 1994).

The subsequent transformation of the economy of the republic has been impressive (Tables 6.1a and 6.1b). Double-digit growth throughout the 1970s was based on new manufacturing industries, oil-refining, ship-

Table 6.1a Singapore: percentage contributions to GDP 1960–93

Industry group	1960	1970	1980	1990	1993
Primary	3.5	2.1	1.2	0.4	0.2
Manufacturing	17.7	24.9	28.8	27.2	25.5
Utilities	1.6	1.7	1.7	1.8	1.7
Construction	5.9	9.9	7.3	5.2	6.7
Trade	27.6	23.3	19.5	17.9	17.5
Transport/Communications	8.5	6.6	10.6	12.2	12.1
Finance/Business	14.8	16.8	19.9	25.0	26.0
Other services	20.4	14.7	11.0	10.3	10.3

Source: Compiled from Department of Statistics, Singapore.

Table 6.1b Singapore: changing economic growth rates (per cent)

	1980	1982	1984	1986	1988	1990	1992	1993
GNP	10.2	6.3	8.5	1.9	12.3	12.6	5.6	8.9
Exports	34.0	0.4	11.2	−2.4	31.2	9.3	1.4	15.6
Imports	33.9	3.4	2.7	3.9	29.0	13.4	2.9	17.1
Tourist arrivals	14.0	4.5	4.8	5.3	13.8	10.2	10.2	7.2
Contribution to GDP:								
Manufacturing	17.0	−5.6	7.4	8.3	17.9	12.4	2.4	9.8
Construction	10.0	35.0	15.4	−22.4	−4.0	16.3	20.0	8.0
Finance/Business	25.0	16.8	12.3	5.9	5.2	19.1	5.3	13.1
Transport/ Communications	13.0	11.0	9.7	8.5	10.6	7.3	8.7	9.6

Source: Department of Statistics, Singapore.

building and repairing and tourism. The second industrial revolution of the 1980s, whilst slow to begin, has eventually produced the 'Switzerland of Asia' with high-tech, high-value industries, an extensive range of producer services and a growing regional financial role (Grice and Drakakis-Smith 1985: 347–59; Grice 1991: 102–18; Rigg 1990; Goh 1992). The great majority of this growth has been financed by the state and foreign investment (Goh 1992; Mirza 1989) but is almost totally dependent on the human resources of Singapore. It is for this reason that the management of its population has been so important to the government and has dominated its policy-making. The nation-state project in Singapore has been firmly structured on the need to ensure not only that the population all pull together in the same direction but also that it presents an attractive inducement for foreign investors.

The consequence has been an attention to human management which is very detailed and has been labelled as authoritarian and intrusive by some critics (Gook 1981: 244–54). In the 1960s and 1970s attention was focused on creating a stable, attractive work force for multinationals. This centred around a strictly enforced population control policy (Saw 1990); an expanded education programme, in which children were divided as early as possible into 'hand' or 'brain' streams; and a massive housing programme designed to extend home ownership to most citizens, giving them a firm interest in maintaining the future security of Singapore. In a parallel process, the People's Action Party (PAP) also sought to undermine political opposition. This it did in two ways: first, by an urban renewal programme that effectively destroyed Chinatown, the physical base of the island's Communist party (it is the rump of this formerly extensive area that has recently been 'conserved'); second, by tightening controls on trade union activities. This latter measure was all the more remarkable as the PAP itself had substantial trade union origins. Essentially, government control was extended over the trade unions so that they became an extension of government itself, rather than representing workers' interests.

Effectively, control over the unions also put a barrier around the development of a proletariat during the period when waged work in factories expanded. Indeed, later government measures have sought to undermine further the development of horizontal class linkages through the adoption of the Japanese model of industrial and corporate relations. By encouraging vertical loyalties within the firm, horizontal solidarity across employment, occupation and income bands has been deflected. A truck driver for Unilever feels more closely bonded in interests and loyalties with a higher-status employee in the same company than with another driver working for General Electric. The emphasis on continuing ethnic and cultural pluralism may be seen as part of the same process of replacing class cleavages, conflict and potential disinvestment, with other foci for group solidarity. Certainly, as far as the government is concerned, this plan has worked well. Multinational confidence and investment in Singapore are high; there has not been a day lost to industrial disputes since 1977.

With the switch to an export-based industrial economy, the 'ethnic balance' of the mid-1960s gave way to favouring ethnic neutrality within the new meritocracy. Singapore likened itself to the Israel of Asia, emphasizing feisty self-reliance and advancement through merit. As part of this process, individual ethnic consciousness was to be submerged into a national identity. English was promoted in schools as a major tool in this process. Ethnicity was depoliticized and reconstituted as cultural identity only. Any attempt to retain a political dimension and to agitate for special treatment was construed as a threat to the stability of the state. It was during this period that politics became the province of an expert elite and organized political opposition of all kinds, not just ethnic, was suppressed. Labour was weakened by the promotion of the pro-government NTUC (National Trades Union Congress) over the more radical STUC (Singapore Trade Union Congress) and SATU (Singapore Association of Trade Unions).

In similar vein the political dimension of ethnic identity was stripped away. Identification with, and support for, external ethnic causes in India, Malaysia or China was actively discouraged. At the same time within Singapore, the complex ethnic mixtures were redefined and consolidated. Thus Mandarin was pushed as *the* language for 'Chinese' to learn, irrespective of individual origins. At the same time, the 'Westernization' of the elite was blurred by a campaign to protect Singapore from the degenerative influences of a morally bankrupt West.

The problem with the meritocracy was that amidst the spectacular economic growth of the 1970s ethnic disparity continued (Chiew and Ko 1991: 116–37) and between 1975 and 1980 the proportion of Malays in the low-income group (earning less than $400 per month (Singapore dollars)) rose slightly to two-thirds of the ethnic community, whereas the proportion of Chinese similarly affected fell by more than half to 42 per cent. As Brown notes, 'meritocracy appeared to be promoting racial inequality rather than equality' (1994: 87). Malay discontent had therefore risen. Ironically so had Chinese disquiet as a result of the promotion of Mandarin- and English-language education. Any open criticism from either community was, however, dealt with very firmly. Around this time, too, the old siege mentality of a small isolated state in a sea of instability had begun to wear thin, particularly with younger Singaporeans. In short, it was perceived that discontent, particularly of an ethnic nature was growing, and at the same time the sense of national identity had waned. Given that the world economic situation was also forcing a reappraisal of the industrial base of Singapore, 1980 saw a restructuring across the board in terms of the directions of development.

The second industrial revolution began in the 1980s, and after a difficult start, has eventually wrought a considerable transformation in the Singapore economy. Down-market, labour-intensive industries have largely been shifted off-shore to the 'inner-triangle' of the Riau Islands and South Johore (Grunsven and Verkoren 1993), where Singapore retains substantial controls through management and technology inputs. At the same time, of course, this process has fragmented any remaining proletarian bonds. Singapore itself now has a new set of knowledge-based industries, producer services and financial institutions, most of which have given it a much higher regional and global profile. Its GNP of US$14,210 per capita in 1991 puts it second only to Japan in Asia and has enabled Singapore to join the first division of high-income states in the league tables of the annual development reports of the World Bank.

In terms of ethnic politics the new emphasis on teamwork, as expressed in the introduction of in-house unions and the adoption of Japanese-style company loyalties, was marked by a substantial effort to re-establish cultural identity. This was not merely to heighten people's awareness of themselves as Chinese, Malay or Indian, but aimed rather to reassert the Asian values of loyalty to family, community and state that would 'counteract the disruptive individualism of western liberalism' (Brown 1994: 88), including the liberation theology of various Christian churches. In similar vein, the potential threat from fundamentalist Islamic developments has been rigorously repressed in favour of the more traditional family-oriented values.

At the same time as ethnic culture has been re-emphasized, however, the pre-eminence of the state as managers and controllers of society has also been stressed, particularly within the context of the very positive expressions of Singapore's 'future greatness', which can only be attained by all working for Team Singapore under the firm direction of the head coach, now Goh Chok Tung. As part of this team development the ethnic groups have been permitted to take up 'approved' interests to develop their own community within the context of the state. Each group has been given official avenues of representation, either directly, in the appointment of parliamentarians, or indirectly through interest groups who 'advise' government. There is a contradiction inherent in this approach, however, in that the creation of an Asian Singaporean identity involves becoming less Chinese or Malay, which contradicts the values that the state is seeking to promote at the community level. Many Singaporeans do not see the government as ethnically neutral and interpret various policies as being in favour of one ethnic group or another. Thus despite the best efforts of government and the real successes it has achieved in creating an ethnically stable, development-oriented society, there seem to be seeds for future instability.

Despite the avowed egalitarian philosophy and the successes in income redistribution, Teo and Ooi suggest that there are still discernible differences between the major ethnic groups in a variety of socio-economic indices (Table 6.2). Occupational characteristics reveal that more than twice as many Chinese are in better paid professional, administrative or managerial positions than Malays, who are more likely to be in production-related work. Not surprisingly, this is reflected in income differentials, with the average Malay monthly income in 1990 being only 70 per cent of that of the Chinese. These discrepancies are made more acute by the larger Malay family size. Such income inequalities translate into social terms even in Singapore, so that, for example, Indian and Chinese families are much more likely to be living in private accommodation than Malays, most of whom live in the excellent but more modest HDB units (Ooi et al. 1993). But not all of the poorest strata in society are Malay, and

Table 6.2 Singapore: selected socio-economic indicators by ethnic group

	Chinese	Malay	Indian
Household income:			
% below S$1500 pm	28.5	35.7	31.1
% above S$5000 pm	17.3	5.6	12.7
Average S$ pm	3213.0	2246.0	2895.0
Average household size	4.2	4.7	4.2
University graduates (%)	5.0	0.6	4.1
Private houses/flats (%)	12.3	1.6	10.6
Employment in (%):			
Professional/technical	17.6	10.0	12.0
Admin/managerial	9.6	0.9	5.5
Clerical/sales	28.0	29.8	25.1
Production-related	38.4	56.0	51.3
Others	6.4	3.3	6.1

Source: Teo and Ooi, Ethnic Differences and Public Policy in Singapore in D. Dwyer and D. Drakakis-Smith (eds) *Ethnicity and Development,* Wiley, London, 1996

not all the Malays are poor. Indeed, because of the overall numerical dominance of the Chinese, most of the poorest households are from this economic group. Moreover, it is, as yet, unclear whether the poorer Malay
households think of themselves as underprivileged in ethnic or class
terms, or both, or neither. No society is completely egalitarian and
Singapore has moved more than most toward this goal, with the Gini co-
efficient in 1990 having narrowed to 0.42. But to what extent do feelings
of relative deprivation undermine the nation-state project upon which
Singapore has focused so firmly, and in what context, ethnicity or class,
are they reflected?

One way in which these questions may be approached is through an
analysis of the changes in population policy in Singapore. Nothing could
be more central to nation-building than having sufficient people (Ministry
of Health 1994), particularly in the case of Singapore, where human resources constitute its only resource. Although ostensibly egalitarian in
terms of their expectations of society, the changing policies in fact prescribed different roles for different groups in society and must be seen as
part of the much broader programme of human resource management
that has been described above (Drakakis-Smith et al. 1993: 152–63). But to
what extent have ethnic pluralism and emerging class interests interacted
with this process? A closer examination reveals that although ethnicity
has been identified by the state as the major threat to stability, nation-
building and development, it appears that emerging class interests could
pose more substantial problems, and have, in fact, elicited a more direct
response from the state.

Population policies in Singapore

When Singapore became an independent state, one of its main demographic problems was not just the ethnic mix but also rapid population
growth, particularly in relation to the availability of employment. Strict
immigration controls were enforced, as was a fertility-control policy
which urged families to 'stop at two'. The family-planning programme
was reinforced by a series of disincentives for those who strayed beyond
two and was linked to a widespread public campaign in which having
more than two children was deemed to be socially irresponsible and
threatening to the stability of the state as a whole. In the context of
ethnicity, all of the major groups traditionally had large families, so the
policy was complementary to the 'neutrality' affected by the government
in ethnic matters. However, as wealthier households already tended to be
smaller, the policy favoured the better-off over the poorer families.

Not surprisingly such campaigning had a substantial impact, as did the
rapidly rising standards of living. Taken together these changes brought
the overall fertility rate down very rapidly indeed from 4.7 in 1965 to
below replacement level by 1977, falling as low as 1.4 in the mid-1980s
(Drakakis-Smith 1994: 43–63). The fall in fertility rates has not been
evenly spread throughout Singapore's population. In particular, there are
differences between the major ethnic groups and also across income/edu-

cational bands. These ethnic and class differentials have apparently attracted quite different responses from the government. The ethnic variation can be seen in Table 6.3 which contrasts earlier data with the situation prior to the introduction of the new population policy in 1987. As is evident, the Chinese population continued to reproduce at well below replacement level, whereas the Indian and Malay populations were just above. These differentials are not likely to narrow in the near future as there are also contrasts in the age at which women marry, with Chinese marrying later than Indians, and Malays marrying earliest. This will obviously affect reproduction within marriage.

Another major variation within fertility data relates to education (and by implication to income). At the same time as overall fertility was falling below reproduction level, it became noticed that an increasing number of educated women were either marrying later or not marrying at all. This situation was not only the result of more women choosing to pursue careers, it was also linked to the fact that graduate men seemed not to prefer graduate women as marriage partners. As a consequence less educated women were reproducing at two to three times the rate of university graduates (Yap 1992: 127–43). This trend has continued through to the 1990s and there has been a steady rise in both the mean age of first marriage and in the number of men and women remaining single (Table 6.3).

The response of the state to these fertility differentials between ethnic

Table 6.3 Singapore: demographic changes 1986–91

	1986	1987	1988	1989	1990	1991	1992
Total fertility rates:							
Total	1.43	1.62	1.96	1.75	1.86	1.77	1.76
Chinese	1.25	1.46	1.84	1.56	1.68	1.58	1.57
Malay	2.21	2.34	2.51	2.62	2.71	2.65	2.63
Indian	1.73	1.76	1.89	1.92	1.95	1.90	1.97
Mean age at first marriage:							
Male	27.9	28.2	28.3	28.5	28.7	28.8	29.1
Female	24.9	25.5	25.6	25.7	25.9	25.9	26.1
Chinese	25.5	25.8	25.8	26.0	26.2	26.3	26.4
Malay	23.8	24.0	24.3	24.2	24.4	24.5	24.6
Indian	24.6	24.9	25.2	25.3	25.4	25.5	25.6
Average age of mother at first birth:							
Total	26.3	27.4	27.5	27.7	27.8	–	–
Chinese	27.0	28.0	28.1	28.3	28.4	–	–
Malay	23.9	24.7	24.8	25.0	25.2	–	–
Indian	25.0	26.1	26.1	26.5	26.4	–	–
Primary ed	25.7	26.4	26.6	26.7	26.5	–	–
Tertiary ed	28.7	29.4	29.3	29.2	29.2	–	–

Source: Ministry of Health (1994), Population Planning Unit, Ministry of Health, annual reports to 1992.

and class groups came much earlier and more intensively for the latter than the former. The reason for this was the concern which had emerged in the early 1980s with the quality of Singapore's work force, the upgrading of which was seen as essential to the success of the second industrial revolution with its switch to knowledge-based industry and producer services. The growing failure of educated people to marry and procreate was considered to be a waste of the pool of talented genes, and was seen by some as a threat to the future economic security of Singapore. The response in the mid-1980s was a series of measures designed to encourage the intelligent to have more children but to retain firm controls on family size for others. Most notable amongst these measures were the sizeable tax incentives given to women with 'superior education' to have more than two children; in contrast similar sums were offered to uneducated women only if they agreed to sterilization after the first child (Saw 1990; Cheung 1990: 35–46 or Drakakis-Smith et al. 1993). At the same time more graduates were encouraged to meet and marry, and presumably take advantage of the new tax incentives, through the government funded Social Development Unit (SDU).

Brown interprets the population changes of the 1980s as part of a package of pro-Chinese policies, 'to learn Mandarin, adopt Confucianism, to reject western values, to have more children' which many perceived as new and unwelcome (Brown 1994: 97). But it seems more useful to interpret such changes in class terms. Effectively, what was happening in the 1980s was, on the one hand, a programme of social engineering with a clear class bias, favouring the better educated in professional or skilled employment and encouraging and developing an elite group solidarity through the SDU. On the other hand, working-class unity was being fragmented by the shift of down-market industry off-shore and discouraged by the extended control of trade unions; meanwhile the numbers of the proletariat were being eroded through accentuated fertility control.

By the late 1980s, however, new concerns had arisen, as it was realized that, due to the continually falling fertility rate, the overall work force would begin to decline early in the next century. The government, at this point, decided to abandon its old 'stop at two' policy and go pro-natal across the whole spectrum of the population. Accordingly in 1987, a new population policy (NPP) was introduced under the slogan 'three or more, if you can afford it' (Saw 1990; Cheung 1990; Drakakis-Smith et al. 1993; Teo and Ooi 1993). A whole range of incentives related to tax rebates, priority in housing and education waiting lists and maternity arrangements were introduced. Although ostensibly aimed at all Singaporeans, some measures still exist, such as compulsory counselling sessions, which encourage the less educated to have few children, whilst many of the other incentives offer greater rewards to the already wealthy, particularly those relating to tax rebates. At the same time there is a commonly held view across all ethnic groups that these new incentives will favour the Chinese more than other ethnic groups, primarily because the Chinese are more dominant in the favoured groups at the 'quality' end of the work force. As the Chinese are so numerically dominant in Singapore and yet exhibit the lowest fertility rate, it is within this ethnic group that any national change must be based. The corollary, it is argued, is that as the Malays are less educated, less numerous and play a less important role in ensuring a secure economic future for Singapore, it would be more responsible for them to at least moderate family size. Brown firmly supports this view, seeing the NPP as one of a series of 'licensed debates' oriented towards re-assuring the Chinese community of its role within Singapore's future de-velopment and encouraging it to become both more Chinese and more Asian (or less Western).

The NPP has, however, provoked less public debate than might have been imagined at its outset. Indeed, at least one survey has revealed that knowledge about the NPP is distinctly modest and that although its per-ceived impact amongst the population is extensive, the actual impact of the NPP on marriage and fertility is quite limited (Drakakis-Smith et al., 1993). Moreover, when evaluated within the context of ethnic and income groups, the latter seem to be much more significant than the former, with the lower and upper income groups reacting in their own fashion to the selectivity of both old and new population policies.

Conclusion

This analysis of the attitudes of Singaporeans towards recent changes in population policy, which have been designed to guide the nation-state project into the next millennium on a sound human resource base, has re-vealed some interesting contradictions. The most important of these is that whilst Singapore is a plural society, ethnicity does not seem to be quite the threat to nation-building for Team Singapore as the government seems to believe or, at least, wishes the public to believe. Certainly, Malays do exhibit different attitudes towards population policy from Chinese or Indians, but many of these differences stem from ignorance rather than dissatisfaction or defiance. The more consistent antipathy to-ward population policies of whatever kind comes from the lower-income group, most of whom are, of course Chinese. This seems to be linked to a lack of education, ignorance of public policy and/or a stronger desire/necessity to have more children. These attitudes, which may or may not be taken as a conscious attempt to subvert the nation-state pro-ject, insofar as it relates to the human resource base, cut across ethnicity and clearly refocus attention on class issues.

It is, therefore, possible to reinterpret many of the human resource management policies in Singapore in quite a different light. Far from in-tending to cope solely with ethnic issues, it could be said that suppression of working-class political solidarity has been the major focus of social pro-grammes. Militant trade unions, for example, have been constantly re-pressed in favour of pro-government organizations and in-house unions which encourage vertical rather than horizontal loyalties. Population policies, too, can be interpreted in this light, with working-class solidarity and influence being fragmented by the continued use of migrant labour and the export of labour-intensive industries. At the same time, the growth of the proletariat is being checked by selective counselling and taxation policies. In contrast, the middle-class is being given every incentive not only to grow rapidly but also to consolidate its power base via the SDU, which has emerged as a very powerful establishment institution.

The general topic of poverty and its impact on class structure in the context of urban Pacific Asia has not been a fashionable area of investigation in recent years (Forbes 1993; Dixon and Henderson 1993: 85–114). The World Bank has made very optimistic noises about the decline of poverty in the region, not unexpectedly attributing its alleged eradication to free-market growth (World Bank 1993; *FEER* 18 February 1993: 66). However, UNDP data reveal quite a different picture of relative deprivation and both the United Nations and World Bank have recently produced major policy statements highlighting growing urban poverty and retargeting many of their policy initiatives towards cities. There is no consistency between economic success and an even distribution of income, with even Singapore itself exhibiting a less than impressive set of statistics, given its economic reputation.

This raises the issue of the nature of complementarity between basic needs provision, human rights and political representation. Several Asian nations, including such diverse political entities as Singapore and China, have good reputations in the former but poor records with regard to the latter. To be sure, there is a vigorous debate on the validity of different Western and Asian interpretations of democracy and civil rights (da Cunha 1993), but it seems more than coincidence that both Singapore and China have operated 'quality control' policies towards their citizens (Kaye 1994: 22). Moreover, within Singapore there exists another group of underprivileged people, i.e. the substantial immigrant population who comprise most of the construction and domestic servant workforce and constitute an underclass as sizeable as those citizens of Singapore who occupy the lowest income echelons. In contrast to the domestic poor, most of these immigrant workers are ethnic Malays (from Malaysia or Indonesia), Thais or Tamils and may serve to fragment the class base even further along ethnic lines.

At present, it would seem that ethno-diversity in Singapore is not the threat to the nation-state that Hettne and others fear it might be in other parts of the developing or developed world. Brown credits this to the successful ethnic policies of the corporatist state in Singapore. It may be, however, that class concerns have played a more important role in human resources management than has hitherto been recognized in this context. Much more of a concern, if not a threat, is the surprisingly persistent deprivation of a more disadvantaged or less educated underclass, local or migrant. A more sustained review of this group of households is not only overdue in Singapore but in the Asian NICs as a whole. In the long term, as Anne Booth (1993) has noted, it may be these social contrasts which are more likely to threaten the stability of the nation-state project in Pacific Asia.

Bibliography

Abrahams, R. (1985) 'Racial and Ethnic Manipulation in Colonial Malaysia', *Race and Ethnic Studies*, **6**(1)

Booth, A. (1993) 'Progress and Poverty in South East Asia', paper presented to the British Pacific Rim Research Group, Liverpool John Moores University

Brown, D. (1994) *The State and Ethnic Politics in Southeast Asia*, Routledge, London

Burja, J. (1992) 'Ethnicity and Class: the Case of East African "Asian"', in T. Allen and A. Thomas (eds) *Poverty and development in the 1990s*, Oxford University Press, Oxford

Cheung, P. P. L. (1990) 'Micro-consequences of Low Fertility in Singapore', *Asia–Pacific Population Journal* **5**(4)

Chiew, S. K. (1991) 'Ethnic Stratification' in S. R. Quah et al. *Social Class in Singapore*, Times Academic Press, Singapore

Chiew, S. K. and **Y. C. Ko** (1991) 'The Economic Dimension', in S. R. Quah et al. *Social Class in Singapore*, Times Academic Press, Singapore

da Cunha L. D. (1993) *Debating Singapore*, Times Academic Press, Singapore

Dixon, C. and **J. Henderson** (1993) 'The Role of the State in the Economic Transformation of East Asia', in C. Dixon and D. W. Drakakis-Smith (eds) *Economic and Social Development in Pacific Asia*, Routledge, London

Drakakis-Smith, D. W. (ed.) (1994) *Economic and Social Development in Pacific Asia*, Routledge, London: 43–63

Drakakis-Smith, D. W., E. K. Graham, P. Teo and **G. L. Ooi** (1993) 'Singapore: Reversing the Demographic Transition to Meet Labour Needs', *Scottish Geographical Magazine*, **109**(3)

Dwyer, D. J. and **J. Eyre** (1993) 'Industrialization in Penang and the New Economic Policy', paper presented to a conference on Ethnicity and Development, Department of Geography, Keele University

Forbes, D. (1993) 'What's in It for Us?' Images of Pacific Asian Development in Chris Dixon and David Drakakis-Smith (eds) *Economic and Social Development in Pacific Asia*, Routledge, London

Furnivall, J. S. (1980) 'Plural Societies', in H. D. Evers (ed.) *Sociology of Southeast Asia*, Oxford University Press, Oxford

Ghaffer, F. A. (1987) 'Regional Inequalities and Development in Peninsular Malaysia 1970–1980', *Malaysian Journal of Tropic Geography*, **16**

Goh, K. S. (1992) 'MNCs Brought Jobs and Sparked Change', *Straits Times Weekly*, 29 August

Gook, A. S. (1981) 'Singapore: a Third World Fascist State', *Journal of Contemporary Asia*, **11**(2)

Grice, K. and **D. W. Drakakis-Smith** (1985) 'The Role of the State in Shaping Development: Two Decades of Economic Growth in Singapore', *Transactions of the Institute of British Geographers*, **10**(3)

Grice, K. (1991) 'Singapore's Successful National Development Policy: the Implications for other Island States', in D. Lockhart and D. W. Drakakis-Smith (eds) *Environmental and Economic Issues in Small Island Development*, Monograph No. 6, Developing Areas Research Group, Institute of British Geographers, London

Grunsven, L. Van and **O. Verkoren** (1993) 'Adjustment and Industrial Change in Southeast Asia', paper presented at the Trilateral Conference on Global Change and Structural Adjustment, University of Amsterdam

Hettne, B. (1993) 'Ethnicity and Development: an Elusive Relationship', *Contemporary South Asia*, **2**(2)

Ho, Wing Meng (1989) 'Value Premises Underlying the Transformation of Singapore', in K. Sandhu and P. Wheatley (eds) *Management of Success: the Moulding of Modern Singapore*, Institute of Southeast Asian Studies, Singapore

Hodder, R. (1992) *West Pacific Rim*, Belhaven, London

Hsiao, M. (1992) 'The Taiwanese Experience', in *Development and Democracy*, Urban Foundation, Johannesburg

Kaye, L. 1994 'Quality Control', *Far Eastern Economic Review*, 13 January

Kong, L. and **B. S. A. Yeoh** (nd) 'Urban Conservation in Singapore: a Survey of State Policies and Popular Attitudes', unpublished manuscript, Department of Geography, National University of Singapore

Lemon, A. (1993) 'Ethnicity and Political Development in South Africa', paper presented at the Fifth Keele Geographical Symposium on Ethnicity and Development, Keele University

Lubeck, P. (1992) 'Malaysian Industrialization, Ethnic Divisions and the NIC Model', in R. Applebaum and J. Henderson (eds) *States and Development in the Pacific Asia Rim*, Sage, London

Ministry of Health, Singapore (1994) *Population and Development Issues*

Mirza, H. (1989) *Multinationals in the Singapore Economy*, Croom Helm, London

Ooi, G. L., S. Siddique and **K. C. Soh** (1993) *The Management of Ethnic Relations in Public Housing Estates*, Institute of Policy Studies, Times Academic Press, Singapore

Quah, S. R., S. K. Chiew, Y. C. Ko and **S. M. C. Lee** (1991) *Social Class in Singapore*, Times Academic Press, Singapore

Rigg, J. (1990) *Southeast Asia,* Unwin-Hyman, London

Rodan, G. (1989) *The Political Economy of Singapore's Industrialization,* Macmillan, London

Saw, S. H. (1990) 'Changes in the fertility policy of Singapore', *Occasional Paper No. 3,* Institute for Policy Studies, Singapore

Seers, D. (1969) 'The Meaning of Development', in D. Lehmann (ed.) *Development Theory: Four Critical Studies,* Frank Cass, London

Teo, P. and **G. L. Ooi** (1993) 'Ethnic Differences and Public Policy in Singapore', paper presented at a conference on Ethnicity and Development, Department of Geography, Keele University

Thomas, A. and **D. Potter** (1992) 'Development, Capitalism and the Nation-state', in T. Allen and A. Thomas (eds) *Poverty and Development in the 1990s,* Oxford University Press, Oxford

Wallerstein, Immanuel (1989) 'The Myrdal Legacy: Racism and Underdevelopment as Dilemmas', *Cooperation and Conflict,* **24**(1)

Williams, C. (1993) 'Ethnic Identity and Language Issues in Development', paper presented at a conference on Ethnicity and Development, Department of Geography, Keele University

Winchester, S. (1991) *The Pacific,* Hutchinson, London

World Bank (1993) *Far Eastern Economic Review* (FEER) 18 February

Yap, H. T. (1992) 'Population policy', in L. Low and M. H. Tow (eds) *Public Policies in Singapore,* Times Academic Press, Singapore

Part 3   Industrialization and labour regulation

The fundamentals of capitalist ethics require that 'you shall earn your bread in sweat' unless you happen to have private means.

Michael Kalecki, 1943

CHAPTER 7

The wage labour nexus in Malaysia's industrial development

Niels Fold and Arne Wangel

Malaysia's industrial development since 1970 has been the subject of numerous studies emphasizing, among other things, the expansion of certain industries, structural barriers on the labour market, accumulation of local (manufacturing) capital, establishment of resource-based industries, state-participation in manufacturing activities and the ethnic configuration of social forces. The studies have produced an extensive and varied knowledge about various aspects of the structural and political transformation process in Malaysia (see overviews as presented by Bowie (1991), Brookfield (1994), Jesudason (1989), Jomo (1986; 1993)). Some of the studies focus on selected elements of the accumulation process but none of them is based on a general theoretical framework aiming at a characterization of different phases within capitalist development.

In this contribution, we adopt the regulation approach as an analytical framework to identify the characteristics of the current phase of Malaysia's industrial development and determine its prospects. Our interest concerns the wage labour nexus. Although this in no way exhausts the analysis of institutional forms in the regulation of Malaysian capitalism, it lies at the core of the growth regime.

The chapter starts out with a brief introduction of the regulation approach. We discuss Robert Boyer's model for institutional analysis as well as his dichotomy of the basic forms of the wage labour nexus, one of four major institutional forms that constitute the growth regime. Then, the regulationist interpretation of developing economies contributed by Alain Lipietz is criticized. We share the point made by Alice Amsden, that Lipietz's rigid typology fails to capture the complexities of development paths in the Third World. However, we maintain the regulationist agenda of linking economic growth patterns to institutional forms and aim at a study sensitive to the wage labour nexus as it unfolds in Malaysia.

Using Boyer's more specific models of growth regimes in Asia and Latin America, defined by the combination of type of wage formation and degree of price elasticity of exports, we discuss the Malaysian case. The competitive wage formation is obvious. However, in explaining why the model argument for a transition from the bastard outward-looking growth regime to one of virtuous export-led growth does not translate into a policy shift, we focus on the genesis of labour productivity. In our concluding effort to outline the specifics of the growth regime in Malaysia, we emphasize a hegemonic political leadership as integral to the relative absence of social contradictions despite the growing differentiation within the industrial work force. We conclude by drawing some methodological implications for the application of the regulation approach to the study of developing countries.

Regulation theory: basic concepts and a methodological approach

The regulation approach is the common name for a large number of Marxist-based, theoretical contributions which aim at a better conceptualization of different phases in capitalist development. The contributions develop concepts and analytical procedures to periodicize capitalist development in the long term while at the same time taking into consideration the implications of time and place specifics. Particular attention is paid to changing forms and mechanisms that secure the maintenance and expansion of capitalism; or, in Marxist terminology, the means by which the expanded reproduction of capital as a social relation is secured. In this sense the regulation approach must be considered as a means of establishing a 'language of the middle ground' that links the structural logic of capital with the many faces of capitalism (Jessop 1990; Walker 1989).

However, despite the common concern for the institutional framework in which capitalist development takes place, there are wide differences between the theoretical contributions. On the basis of a comprehensive review, Jessop (1990) identifies certain 'schools' each having a common theoretical point of departure and focusing on the same fields of regulation. But owing to the lack of a strictly coherent theoretical framework (even within some of the 'schools') as well as to the changing focus of individual scholars there are uncertain distinctions between them and mutual overlapping.

This chapter focuses on selected works of two of the influential (French) regulation theorists, namely Alain Lipietz and Robert Boyer. Even though there are differences between these two writers they can be considered as formative and representative of one 'school' (Kotz 1990). In the remaining part we shall designate this narrow version as 'regulation theory'.[1]

The two central concepts within regulation theory are the regime of accumulation and the mode of regulation (Lipietz 1986). A **regime of accumulation** denotes a specific and stable allocation of the net product between consumption and accumulation so that circulation of capital at the aggregate level is ensured. According to Lipietz it is possible to formally represent a specific regime of accumulation in a coherent schema of reproduction, thus demonstrating that aggregate production equals aggregate consumption under a given social allocation of resources.

However, this is only the necessary condition; there are no organizing rules that oblige individual capital and social agents within the system to behave in accordance with the structure and thereby support the sustainability of the regime. The actual existence of a regime of accumulation requires a specific mode of regulation. The mode of regulation constitutes a body of interiorized rules and social processes (norms, habits, laws, regulating network, etc.) that ensures the stability and dynamics of a regime of accumulation. Thus, a specific mode of regulation 'fits' a specific regime of accumulation.

This is not yet another functionalist construction within the social sciences. The mode of regulation should not be perceived as an invention geared to make the regime of accumulation work. Within the regulation approach it is stressed that the specific combination of a regime of accumulation and a mode of regulation is the outcome of social conflict. A stable combination of the two will condense under certain conditions and in certain periods where social relations are reproduced without crisis. At the same time, this means that social contradictions, obvious or hidden, embodied in the regime of accumulation, but previously apprehended within the mode of regulation, for one or the other reason, may materialize in social struggles and destabilize the regime of accumulation. This new round of struggles may, under certain circumstances, lead to a change in the pattern of social and economic reproduction of the system, i.e. the establishment of a new regime of accumulation and mode of regulation.

There is a large gap between the basic concepts of regulation theory (i.e. the regime of accumulation and the mode of regulation) and empirical research. The theoretical framework encompasses a number of different elements in the mode of regulation. Religious, legal and cultural norms and rules are materialized in different kinds of institutions and act in combination with economic institutions to maintain and possibly expand the regime of accumulation. Viewed in this perspective, the regulation approach is extremely complex in terms of constitutive concepts and relations, not to mention the necessary methodological elaborations for empirical research.

Boyer (1994) has made an effort to outline a methodology for institutional analysis within the framework of regulation theory. He identifies four institutional forms that determine whether the inherent conflicts and disequilibria in capitalist accumulation are encapsulated. The institutional forms comprise the wage labour nexus, the forms of competition, the state configuration, and the relations with the international regime. The national economic system is moved into a state of sustained growth, a so-called growth regime, if the four institutional forms propel a productivity regime and a demand regime that *ex post* are coherent. The growth regime is characterized by being self-equilibrating with respect to internal dynamics as well as external shocks.

As such, the model is a methodological direction to a study of the mode of regulation, as it points towards four institutional forms that influence the allocation of the net product between consumption and accumulation, i.e. the regime of accumulation. As shown in Figure 7.1, each of the institutional forms exerts an influence on both the productivity regime and the

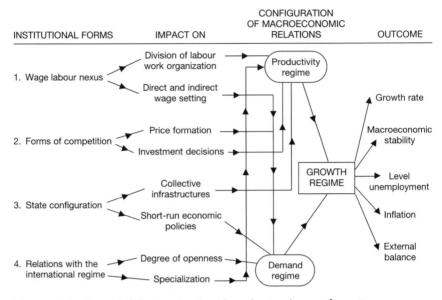

INSTITUTIONAL FORMS   IMPACT ON   CONFIGURATION OF MACROECONOMIC RELATIONS   OUTCOME

Figure 7.1 A model for institutional analysis of growth regimes
Source: Boyer (1994)

demand regime via its impact on certain subcategories. The subcategories are grouped in pairs, each linked to either the productivity regime or the demand regime.

There are, however, a number of uncertainties in the model, primarily in relation to the rudimentary explanation of the substance of the four institutional forms, their impact on the subcategories belonging to each of them, and their mutual relationship. As regards the last issue, the wage labour nexus is implicitly determined as the key institutional form. According to Boyer: 'The wage labour nexus describes the configuration associated to a given state of division of labour, as well as income distribution ... Each stage in the history of the division of labour is associated with definite factors of productivity increases, which combine in various proportions the impact of specialization, learning by doing, design of equipment, the size of minimum efficiency scale' (Boyer 1994: 38).

Boyer summarizes his discussion on variations of the **wage labour nexus** by identifying two main visions of the institutional form. The first one is based on low wages and external defensive flexibility towards unexpected disturbances; in this case, a fair performance might be achieved in the short run, but at the expense of rising social inequality and barriers to long-term organizational and technical change. In the second one, savings on labour costs are replaced by the capture of key innovations and employment of polyvalent workers in a regime of internal offensive flexibility towards surviving international competition; this approach might involve short-run rigidities, but creates pressures for long-run innovation, rising wages and social peace (Boyer 1994: 61).

In a later section we shall return to these basic types of flexibility as they occur in the Malaysian context. First we discuss the approach and the methodology outlined for a regulationist analysis of growth regimes in developing countries.

The regulation approach applied to growth regimes in the South

Lipietz (1986) has also applied the regulation approach to a categorization of industrialization models in developing countries. New concepts are introduced such as 'sub-Fordism', 'bloody Taylorism' and 'peripheral-Fordism', indicating the heavy theoretical ballast from the discussion centred on the industrialized countries.[2] However, this tri-polar categorization does not enrich the theoretical knowledge. The new terms are merely synonyms for well-established concepts such as import-substituting industrialization and different phases of export-promoted industrialization, i.e. the establishment of labour-intensive production of consumer goods or components for export markets (primarily textiles, clothing and electronics) and the additional production of durable consumer goods (including cars) both for export markets and for the domestic market.

Amsden (1990, 1994) points out that the concepts prepare the ground for misinterpretation of the development dynamics in the Third World. The industrialization process, particularly in the East Asian high-growth economies, cannot be interpreted as a diffusion of industrialized-country capitalism. The industrial expansion in the region has to be conceptualized as a qualitatively new process based on a specific and complex interaction between state and market mechanisms as well as a new way of indigenous absorption of technology, a process of so-called late industrialization.

Amsden's criticism is right concerning the unsuccessful attempt to modulate the concept of Fordism on Third World industrial development, but the criticism does not get to grips with the basic concepts and relations within regulation theory. We think that there is a need for further elaboration of regulation theory and that the approach has the potential to develop concepts and analytical procedures that are adequate for analysis of processes at a level between the structural logic of capitalism and the many different historical forms of capitalism.

In this perspective, Lipietz's tri-polar categorization is indeed rather disappointing, as the introductory sections of his paper point to a framework for analysis that adopts an open attitude towards variations in the capitalist development process of developing countries. According to this view, an outline of the regime of accumulation is principally produced by internal class struggle and consolidated by forms of regulation sustained by the local state. Thus, the state is seen as the archetypal form of any regulation within a social formation. It is within this institutional form that periodic compromises between different and opposing social groups are entered, moulded and regulated, stifling endless struggles and ensuring a (periodic) crisis-free capital accumulation. The national social formation is proclaimed as the basic analytical starting point and not as some all-embracing concept of global capitalism. External relations might play decisive roles in the dynamics of some existing regimes of accumulation, but that does not imply that they were initially established for that purpose. By way of a methodological conclusion, he states, 'Thus the goal is to study for itself each national social formation, to observe the succession of its regimes of accumulation and modes of regulation, to analyze its expansion, its crisis and the role therein of its external relations' (Lipietz 1986: 21).

Despite its very general content, this short statement opens up a much broader conceptualization of the differentiated capitalist accumulation process in developing countries. At the same time it provides a starting point for methodological work related to analyses of different aspects within the development process.

In his application of regulation theory to developing countries, Boyer (1994) also singles out the wage labour nexus as the institutional form of overall importance, but a number of specifics with respect to the relation between labour institutions and the growth regimes are stressed. On the basis of studies from Latin America four important features of labour institutions in developing countries are underlined:

1. The institutions are crucially shaped by the type of integration of the national economy in the international economy.
2. The technological dependence stemming from this integration may result in a demand for unskilled labour that restricts the institutionalization of the labour market.
3. The agricultural sector heavily influences the functioning of the labour markets.
4. The effects of the informal sector on the formal labour market must be taken into account.

These features are well known and important for the economic development process of developing countries. As such they need to be considered in relation to an expansion of the sphere of application for regulation theory. It is, however, difficult to find the theoretical cogency in these additions to the original analytical model. The integration in the international economy is already included in the model as one of the four institutional forms (see above), thus Boyer (implicitly) establishes a relationship between two of the institutional forms, a relationship which he did not comment on in the previous presentation of the analytical model. Moreover, the technological dependence is not an 'independent' feature but closely linked to the type of international integration of the industrial sector, i.e. to the form of the industrialization process. Lastly, the influence of the agricultural and informal sector on the (industrial, formal) labour markets is not formalized or elaborated in the remaining part of the paper, even though it is argued that these features must be integrated in regulation theory. Clearly, there are many things to take up in order to improve the theoretical and analytical approach.

What is of particular interest in relation to our present purpose, however, is Boyer's subsequent outline of a scheme which is based on a theoretical model of an outward-oriented East Asian economy.[3] It refers to four different growth regimes, each one a combination of the relationship between the type of wage formation (competitive or institutionalized) and the price elasticity of exports (high or low) (see Figure 7.2).

The 'bastard outward-looking growth' regime is characterized by highly competitive wages and high price elasticity of exports. In a situation of booming world market demand, rising real wages might cause a decline in world market shares. Contrary to this regime, some kind of wage contracts linked to productivity increases are established in the 'virtuous export-led growth' regime. This regime, in essence, is similar to Fordist accumulation, and a boom in world market demand would result in growing productivity and output.

Two other regimes are included in the scheme; they are both characterized by low price elasticity and importance of exports. In the 'stagnationist' regime, growth is retarded by competitive wages whereas the 'Fordist growth' regime offers a better result (sic) (Boyer 1994: 81). According to Boyer, the straightforward conclusion of these model-based considerations is that the result:

> ... totally challenge[s] the conventional view according to which export-led growth supposes strictly competitive wages. This might have been the case in the early phase of the process, but it is no longer an efficient strategy when the economy is hitting a full employment barrier and/or rising social demands for economic and political rights. Again, each regime exhibits a strong historicity: what was good during one period might become detrimental in another phase of the same process. (Boyer 1994: 82)

Boyer gets carried away by his efforts to test the results of the theoretical model in quantitative terms at the regional level, i.e. Latin American

Exports \ Wage formation	Competitive	Institutionalized
Highly sensitive to price	BASTARD OUTWARD LOOKING REGIME: Trade-off between growth and productivity	VIRTUOUS EXPORT LED GROWTH: Cumulative growth and productivity increases
Low elasticity and importance of exports	STAGNATIONIST REGIME: Inability to benefit from world growth	FORDIST GROWTH REGIME: Wage increases might benefit both productivity and growth

Figure 7.2 A typology of outward-oriented growth regimes
Source: Boyer (1994)

versus Asian economies. As expected, the results are very general and all concern for empirical sensitivity is lost. Nevertheless, Boyer has sketched an analytical framework which we think has the potential for further elaboration in a contextualized framework.

The wage labour nexus in Malaysia: the debate on forms of flexibility

Boyer's two general categories of a wage labour nexus are to some extent reflected in the debate on labour market policies between the Malaysian government and its critics. Checking wage increases is a principal policy concern for the Malaysian government in order to fight back inflationary pressures and maintain labour cost as a key incentive for foreign investment; the government claims that the wage trend is overtaking the rate of growth in productivity. Nominal wages are rising in Malaysia; from January to July 1995 wages in the manufacturing sector increased by 9.7 per cent (compared with 6.7 per cent during the same period in 1994) (Ministry of Finance 1995).

The use of foreign labour is the government's principal measure to maintain a competitive labour market. An estimated more than one million Indonesian workers are present in Malaysia, first of all in the construction sector. They are replacing Malaysian labour in the plantations, and more recently, together with newly arrived workers from Bangladesh, Thailand and East Malaysia, they are filling vacancies in the manufacturing sector. Labour contractors provide this work force with two-year contracts; this is a most welcome arrangement to employers, who have to cope with exorbitant turnover rates (see Ministry of Finance 1994: 171ff.; Human Resources Development Council, 1994). Furthermore, employers have launched and relaunched the concept of flexi-wage on several occasions with government backing. Flexi-wage ties wage levels to up- and downward trends in company performance. Unions have rigidly opposed the idea; they have also met with reluctance from employers when enquiring about detailed financial information as the baseline for such wage-setting in the individual company.

By contrast, the demands of the union movement are a national minimum wage by legal enactment, which is flatly rejected by the government, and shorter collective agreements. The trade union movement argues that the pay rise is a matter of catching up with earlier losses of purchasing power, while at the same time matching the increasing rate of inflation, about 3.5–4 per cent according to the official consumer price index (Ministry of Finance 1995).

Standing (1991) observes that the Malaysian labour market displays minimum regulation to counter efforts of the companies to achieve a high external flexibility; wage and non-wage costs are contained by the companies and possible 'rigidities' affect the nature of employment far more than the level of employment. Standing rejects wage-cutting as a measure to maintain competitiveness on the world market and defines the developmental challenge for Malaysia as a definite departure from static flexibility. In his paper, Standing goes to great lengths to argue important advantages of employment security regulations in terms of increase in productivity, lower turnover and higher work flexibility. These elements are considered to be necessary for the transition of the Malaysian economy to an economy based on dynamic efficiency and technical skills rather than the present one based on low-cost, semi-skilled labour (Standing 1991: 234).

The Malaysian government is also seriously concerned about the supply of a work force with adequate skills for the manufacturing sector, whether it is a question of skilled workers, technicians or engineers. To avoid bottlenecks, sufficient manpower is considered most crucial in order to sustain high economic growth in Malaysia. In the competition for foreign investment, Malaysia must be able to move up in terms of providing manpower for the operation of sophisticated manufacturing technology, as simple, labour-intensive production is shifted to Thailand, Indonesia, Vietnam and China (Pillai 1994).

At present, the combined effort of public educational institutions, private training courses and skills-upgrading centres organized by subsidiaries of transnational companies in Malaysia falls short of fulfilling the needs for skilled staff. The basic in-house training that major companies undertake is hampered by extensive job-hopping in a situation of close to full employment.

Thus, for transnational companies depending on rapid model change towards higher-end products for export, even basic training is a constant effort at sharply rising cost. Other companies, including Malaysian-owned medium-scale companies, experience negative incentives in a highly competitive labour market with regard to skills development. As argued in general terms by Boyer: 'If ... the labour contract is essentially a spot transaction or at least is of short duration, firms will under-invest in the specific skills of workers and will prefer layoff to internal flexibility, employment reduction to product innovation' (Boyer 1994: 54).

Standing (1991) also observes that the government considers the particular skills requirements of the foreign-owned manufacturing sector as an independent variable, which the educational profile of the Malaysian work force should match. In his opinion, this kind of passive response, for example to a change of product models, represents another shortcoming in the labour market policies:

> 'Too often, a perceived shortage of skilled labour is presumed to mean that the appropriate policy is more schooling or training. It is conceivable that a more appropriate and cost-effective approach would be a policy to alter job structures rather than the attributes of people required to fill them. It is technologically deterministic to focus exclusively on training'. (Standing 1991: 215)

To sum up: Boyer made a basic distinction between a wage labour nexus characterized by external defensive flexibility and internal offensive flexibility. Guy Standing's encounter with Malaysian politics revealed his preference for dynamic efficiency rather than static flexibility, as he terms the contrast between the two growth regimes. However, the discourse on models fails to explain why such arguments of economic rationality pointing to specific conditions for improved growth performance do not translate into a shift in governmental policies. This critique still applies when Boyer specifies a model in relation to Asian countries as a juxtaposition of a bastard outward-looking regime and virtuous export-led growth.

In our opinion, however, Boyer's more elaborate model has the advantage of providing a match for the Malaysian case. The model carries the analysis forward by asking questions about the mode of wage formation and the development of productivity. Thus, in the final part of our contribution, we will take a closer look at the current competitive mode of wage formation in Malaysia and the conditions for increasing labour productivity.[4] In the concluding outline of the growth regime in Malaysia, we offer an answer to the question of why the Malaysian government is 'stuck' with a seemingly 'inadequate' growth regime. This chapter ends by proposing a wider perspective on the social differentiation produced in the growth process and the political containment of the evolving contradictions.

Wage formation

The legal framework regulating the labour market itself indicates a highly competitive regime. Conditions of employment are subject to minimal limitations. The law does not stipulate a minimum wage or any specific regulation on wage-setting, apart from the calculation of rates for overtime and work on public holidays. Furthermore, the absence of a public welfare system implies that any discontinuity in workers' wage income due to illness, loss of employment or unfitness to work has to be compensated for with whatever resources are or are not available in the family household.

The system of labour laws has eliminated any political function of workers' representatives in the area of labour policies. Furthermore, it is geared towards weakening the economic function of collective organizing. Deliberate fragmentation and a *de facto* ban on industrial action impedes trade unions in the defence of members' interests and achieved standards (Wangel 1991).

The explicit goal of the government is to control the current pressure for higher wages. As the labour market is getting close to a situation of full employment, the antidote is considered to be a more intense competition to keep labour costs at a relatively low level (Rasiah 1995).

In some production sectors there is a formalized collective agreement regulating the relationship between employers and employees. As an example, various industries within metal fabrication have been organized by national unions, achieving a membership of close to 100 per cent in the plants covered, and about 40 per cent of the total number of employees in that particular industry. At national level, an average of about 10 per cent of the working population are members of trade unions. For organized workers, wage setting is regulated by a collective agreement, which covers almost all employees in a particular plant and has a duration of three years. This agreement defines the basic monthly wage according to job grade and seniority, as well as annual increments, annual bonus and additional non-wage benefits.

The contractual relationship between employers and employees does not develop into an institutionalized cooperation, whereby the two parties engage in a stable compromise on productivity growth and wage and employment conditions. On the contrary, in particular at the point of renegotiation, the fragility of the relationship between the two parties is evident. As an agreement is about to expire, the employer might refuse to launch negotiations early on the union's proposal for a new agreement. He might frustrate the union by postponing meetings. Finally, the employer can stall completely on the union's wage proposals. By this action, the employer deliberately seeks the intervention of the authorities. The fixing of wages is then settled either by a sort of 'compulsory' arbitration, as officers from the Ministry of Human Resources step in, or through a case brought before the Industrial Court. By the latter course of action, the union side is exposed to two major disadvantages. First, a substantial backlog of unresolved cases implies that a new collective agreement will take effect only a couple of years down the road. Second, the decisions of the Industrial Court on wage increases relate closely to the consumer price index (CPI), which is much criticized for not reflecting the actual inflation rate.

Even a normal renegotiation takes place only every three years. Thus, adjustments are still delayed, although trade unions are currently more successful in their role of securing above-average wage rates for their members as they ride on a wave of upward wage pressure in a tight labour market.

The legal framework supports a competitive labour market, which in turn creates structural conditions that further weaken the bargaining position of unions. As an example, the drive for uniform collective agreements within an industry may be hampered by various restrictions on the union. Newly organized firms are given a 'period of grace' before being expected to adapt fully to the average wage levels in the organized part of the industry. This is part of a trade-off with the employer to gain his recognition of the union. Special consideration is also given to firms experiencing financial difficulties, as job security for the members is first priority.

A recent conspicuous characteristic of collective bargaining is the fact that improved results for permanent workers have been achieved at the expense of a previously non-negotiable union ban on casual and immigrant workers in organized companies. The undercutting of non-skilled workers by illegal as well as government-sponsored imports of foreign labour was mentioned above. This is one major factor affecting the labour market position of unionized as well as non-unionized workers. Although the authorities have followed a stop–go policy on foreign labour due to evident irregularities and public concern about social implications of the influx, the supply of this low-paid, bonded labour is vital to the survival of a number of small and medium-scale companies.

At the same time, the labour reserve in Malaysia is itself far from exhausted. In spite of extensive labour migration, rural areas still hold large labour pools, which are gradually being mobilized by companies relocating their lower-end production to those areas, or through improved infrastructure, which makes industrial centres more accessible. These new workers enter the labour market, as their colleagues did earlier, as a semi-proletarized work force. One example of employers' haste in scooping up all available manpower is the undergraduate students who are employed on a short-term basis during university vacations.

Making use of intermediaries is another aggressive tactic in recruitment. Some employers make use of company drivers as labour contractors. Workers are provided with transport by their employer in order to reach work from scattered villages and at the odd hours required by shift work. The drivers of the factory vans may be offered better terms by another factory if they manage to divert their load. These workers are then confronted with the choice of either accepting a new place of employment or losing their transport.

Wage formation is influenced by movements of supply, but the type of remuneration offered by employers and the way workers respond to these offers also affect the trend of wages. In Malaysia, wage income is composed of a basic monthly salary (extensive use of piece-rates can only be found in the garment industry) and variable benefits. Job offers relate to both components.

For some workers, job-hopping can be triggered by marginal differences in the basic monthly pay. The arrival of new investors or a significant expansion of an existing plant spark off a wave of job movements. The new or expanded factories will poach experienced staff to start up their own operation while at the same time causing disruption and increased training costs to others. Earlier agreements among personnel managers to maintain similar wage rates and to avoid comparative advertising have evaporated in the course of stiff competition for labour.

Other workers focus on the package of variable benefits, e.g. annual bonus of one, two or even four months' basic pay, free medical treatment etc. Transnational companies have the edge in these offers, with the added advantage of their well-established brand name inducing a sense of job security.

The working environment and social relations represent additional non-wage factors in job-hopping. Moving from the tedious low-paid sewing work in a garment factory to what is perceived as trendy high-tech work in the air-conditioned environment of an electronics factory is one such motivation. Friendship groups take up employment together, support the individual operator in coping with monotonous assembly-line work, and also become a key factor in decisions on shifting workplace.

In conclusion, one can observe that wage formation in Malaysia's industries takes place in a frenzied exchange of labour. The labour market is partly institutionalized by collective agreements, but union recognition, renegotiation of agreements and solving of disputes represent aspects of the relationship between employer and organized workers which may, at any point, turn it into one of confrontation. The labour market is non-transparent in the sense that each actor has to follow his or her own leads as to the best ways of optimizing one's position. The lack of precise labour market information means that wage trends are only sporadically guided by real movements of supply and demand. Adding to the flux of the labour market is the expansion of informal and casual employment. An individualized relationship between employer and workers presents a barrier to unionization and further adds to the trend of disparate wage-setting. A restrictive legal framework for union activities constitutes a fixed barrier restraining collective, defensive measures. Employers are free to address individual motivations, as they put together the structure of incentives in the pay package.

Thus, in the wage formation on the labour market, labour is not in a position to make full use of the long-term upward trend of the basic wage caused by labour shortages. The labour market is competitive to the extent that different groups of workers are pitched against each other. Segmentation and polarization persist. Some of the gain comes in the form of variable benefits, which employers may easily reduce, once a period of economic downturn starts. The government, for its part, is committed to a business climate in which low-cost labour is readily available, possibly in the form of imported labour.

Potentials for increase in labour productivity

The process of moving from Boyer's bastard outward-looking growth regime to one of virtuous export-led growth includes a significant increase in productivity. The simultaneous transition of the wage labour nexus from a competitive to an institutionalized mode implies that productivity gains can be fully utilized as stimuli for further growth. What is the current basis for the development of labour productivity in Malaysian manufacturing and where does the potential lie for a substantial increase? This section aims to provide some of the answers to these questions.

The changing role assigned to subsidiaries in Malaysia in the global strategy of transnational corporations has brought in higher-quality products, automation technology and more elaborate managerial styles (Lim and Pang Eng Fong 1991). However, in terms of work organization there is no indication of flexible specialization operated by polyvalent workers; on the contrary, mass production based on fragmented work tasks continues.

In Amsden's argument, an industrialization process based on borrowed technology is entirely different from one based on the generation of new products or processes. She identifies one main characteristic in the work organization, namely the continuous troubleshooting carried out by all workers on the production lines: 'The shop floor tends to be the strategic focus of firms that compete on the basis of making borrowed technology work ... where the achievement of incremental, yet cumulative, improvements in productivity and product specification occur, and therefore enhance competitiveness' (Amsden 1990: 17).

Methods of quality control and problem solving exercised in small group activities extract higher productivity from workers within a limited agenda of employee participation focused on individualized labour-management relations, which sidelines or makes obsolete any trade union representation.

However, the incremental accumulation of troubleshooting develops operational capabilities only. In the Malaysian case, even this type of experience barely contributes to the development of a national technological capacity. There are very few Malaysian industrial capitalists who can borrow foreign technology in the first place. Advanced technologies are present in Malaysia due to direct investments by foreign transnational companies. The example of the largest Japanese investor in Malaysia, Matsushita, shows that operational experience is relayed back to headquarters in Japan, which maintains control of the core technologies (Wangel 1993).

The developments in the Matsushita Airconditioner plant are part of a broader trend. The earlier phase of import-substitution did not require advanced machinery to produce goods targeted at the domestic market only. However, following the appreciation of the yen after 1985, the inflow of direct investment with the objective of relocating an export base needs production facilities that can manufacture quality products to the level of international competition. These new operations are equipped with automatic and robot-controlled, high-tech inspection machines, and they import parts from Japan and the NICs in order to fill immediately the gap in the technological level between the Japanese and local factories.

Yamashita (1991: 19) points out that this gap between the Japanese requirements and the Malaysian technicians makes these operations 'inactive in transferring higher technology'. A 'black box' of production technology is created, as the gap is filled by automated and computer-controlled machines, which are not matched by an up-skilling of the work force, because that is considered to be too lengthy an exercise. This knowledge then remains inaccessible to Malaysian technicians. These Japanese subsidiaries become strong exporters of high-tech products, but rely on imported components, automatic assembly equipment and Japanese-style management.

Preliminary evidence from the electronics industry, including the disk drive manufacturers who have recently moved from Singapore to Malaysia, seems to indicate that advanced technologies are applied selectively in the production process. The number of production operators going for training at the main site of transnational companies is relatively small compared with the total work force in these plants, which are considered 'high-tech'. Rather than flexible specialization, the work organization is characterized by computer-assisted Taylorism. Automation is introduced in certain areas of the assembly line to speed up the flow, to avoid bottlenecks and to enhance product quality; e.g. test functions are supported by a computerized diagnostic system, which suggests corrective procedures to the production operator. In addition, materials handling and output performance are made transparent by a computerized information system so that effort can instantaneously be adjusted in the light of fixed production targets.

The kind of skills development achieved is reflected in the fact that production operators serve as trainers when a transnational company decides to relocate a lower-end product from Penang to a new plant in China. A complete production line of Mandarin-speaking operators is transferred for a couple of months to the operation in China and is responsible for on the job training of their Chinese colleagues (interview with Managing Director, Motorola (M) Sdn.Bhd., Penang, April 1994). By this transfer of experience, workers assist the companies in exploiting international wage differentials – an exercise which at the same time could erode their own employment base.

Skills upgrading for production operators occurs as the manufacturing process is changed due to introduction of new product lines. Part of the process might be automated to facilitate complicated assembly and higher quality standards in more sophisticated products. Thus, the skills upgrading concerns operational capabilities related to certain products and does not necessarily develop general qualifications, which could enhance the status of the worker on the industrial labour market. A few exceptional companies do provide training for male workers equivalent to certificate craft-related courses which are nationally and even internationally recognized. The female production operators of the nineties are exposed to advanced work procedures which are different from the simple, manual assembly performed by their colleagues twenty years earlier. This handling of computerized machinery and test equipment requires new skills, but the Taylorist division of conception and execution persist.

These observations reflect a gap between the technologies available to the subsidiaries of foreign transnationals and the present capacity to absorb the knowledge necessary for Malaysian workers and technicians to master advanced operations. Increased labour productivity can still be achieved at the conveyor-belts manufacturing long series of standard products. But the removal of bottlenecks and other measures to speed up the production flow is far from the kind of cooperation between multiskilled workers in semiautonomous workgroups, which has the potential for innovative change in processes and products.

The rationale of the 'irrational' growth regime

Why does the Malaysian government maintain coercive regulations *vis à vis* labour in an attempt to suppress the effects of a competitive wage regime rather than transforming it? Why does the government hesitate to contractualize the wage–labour nexus to secure an optimal increase in productivity, as some kind of agreement between employer and employees on sharing the gains is established? And why does the social differentiation produced by a wage labour nexus in the competitive mode not evolve into open, political conflicts? We believe that the answers to these questions are interrelated and that together they constitute the outline of an explanation as to why a transition to virtuous export-led growth is stalling.

First, it is doubtful whether the Malaysian economy is in a position to embark upon rapid productivity growth. For several reasons a radical shift in the short term is not viable. Levels of labour productivity which compare favourably to those of industrialized countries can be found in subsidiaries of transnational companies. But a number of structural factors inhibit the development of a national technological capacity, which is necessary for continuous progression involving the adoption of radically new production concepts of computer-integrated manufacturing and in-house innovation in materials and processes.

Within the resource-based industries a national innovative base has been developed (see Fold 1993), while manufacturing depends upon technology imports organized by foreign investors. The government has provided various economic incentives to promote technology transfer and research and development activities in foreign subsidiaries. However, for these incentives to take optimal effect a technological infrastructure is necessary, among other things, and this is not yet in place.

The successful export-oriented industrialization in terms of sustained high economic growth has produced an industrial structure, the range of which is too narrow to form a diversified and integrated industrial base. Within the export enclaves, the foreign subsidiaries have created local off-shoots or 'imported' their sub-contractors (Rasiah 1994). Their success has absorbed the pool of skilled labour, creating demands on the system for vocational training way above its capacity. A shift to a growth regime based upon rapid and continuous productivity increase is confronted with absolute skills shortages. Faced with these structural weaknesses, which can only be overcome in the long term, the government does not intervene to abandon altogether the traditional source of growth provided by labour-intensive, lower-end production, as happened in Singapore.

Second, since 1970, Malaysian politics has explicitly been configured on the redistribution of economic resources among the three ethnic groups. Sustained growth has become a political imperative, as the ethnic compromise requires improvement in living standards for all groups. Growth itself rather than its source is important. Currently, workers experience positive adjustments of their nominal wages. However, an unreliable consumer price index makes the estimate of any change in real purchasing power inaccurate. Some workers have speculated with their personal savings in the long-term bull-run on the stock market. Substantial gains have also been made due to another effect of the close-to-overheated economy, i.e. a steep climb of property prices in urban areas. Almost all groups of workers experience some increase in their available income, which is reflected in more widespread consumerist lifestyles. In political terms, this development creates a sense of comfortable progress common to groups among whom differences in living standards are widening.

The frenzied exchange of labour in Malaysia's industrial centres has developed a segmented and polarized labour market as employers seek recruits to honour their export orders and workers search for opportunities and new lifestyles. In a rough attempt to describe the structure of the industrial workforce, one can identify at least four main groupings:

1. The original migrants from rural areas, who became a semi-proletariat occupying manual, non-skilled jobs for a limited number of years in their process of social mobility
2. An increasing number of workers without skills, who are employed temporarily on the fringes of the labour market, as employers increase external flexibility and make use of foreign labour
3. Those workers who are covered by a collective agreement, but whose position is challenged by casual labour and younger colleagues who relate more to individual recognition
4. The technicians, who have been professionally trained for specialist functions and senior workers upgraded to handle more sophisticated operations, i.e. 'core workers'.

Members of these groups do not constitute individual economic units. They belong to a household forming an extended family. Thus, the differentiation between them in terms of the level of wage income represents only one among several components of economic resources.

The economic strategies of the households of workers in various groups differ significantly. Labour migration represents a classic supplementary source to the family income. Urbanization provides another set of economic options, just as transformation of rural areas due to the penetration of infrastructural development may change economic strategies of local landowners altogether. Workers in higher-end transnational companies feel that job security and working conditions are no longer a problem and focus on issues of personal relations. Differences in political outlook tend to widen too. For example, the skilled core workers find themselves in an ambiguous class position when they are promoted to middle-management positions.

However, the growing inequalities do not lead to outspoken social contradictions and class formation. Regulatory barriers for combination and articulation of interests in the labour market remain an important explanatory factor in understanding the docile political climate. We think that the 'addiction' to economic growth has become a significant ideological factor. It was originally constituted as an essential condition for the ethnic compromise, then followed by the experience of achieved long-term improvement of living standards across the stratified segments of society, and recently restated in a developmental vision, 'Vision 2020', specifying the prospects of future wealth.

In the perception of Malaysia's political elites, Vision 2020 comprises different aspects. Whilst aiming at full economic, politic, social and spiritual development for Malaysia in 30 years, it is also intended to lead to national unity among the people, who will come to identify themselves as Malaysians (Ahmad Sarji Abdul Hamid 1993). It is a modernist development vision based on Western rationality but attempting to distinguish 'Malaysian values' from those of the West (Kahn 1994: 24).

No opposition group has yet been able to formulate an alternative policy achieving broad-based support. When the elites define Vision 2020 as the route to achieving the status of an industrialized nation, they make explicit reference to current political stability. One shared key notion is that the confidence of foreign investors in the ruling coalition is vital to the realization of the project.

Conclusion

The regulation theory – in the rather narrow version presented in this chapter – offers a useful conceptual framework for analysis of the institutional relations behind the industrial dynamics in a particular national context. It is, however, a precondition for a fruitful application that the analysis goes beyond the stigmatized conception of Fordism as the crucial growth regime and seeks other ways to explain growth paths than in terms of variations or differences from Fordist institutions. This is particularly relevant in the case of social formations that have started to industrialize under different internal and external conditions from those experienced by the industrial countries in the North.

A recent effort to establish a model for institutional analysis in line with the regulation theory was taken as a point of departure. Implicitly, the wage labour nexus was singled out as the most important institutional form within this analytical approach. Two different forms of the wage labour nexus, characterized by either external defensive flexibility or internal offensive flexibility, were outlined. The difference basically boils down to the presence or absence of various institutions for labour market regulation.

The same conceptual dichotomy of the wage labour nexus as described above – here in terms of static flexibility and dynamic efficiency – exists in the Malaysian political debate. Major issues are the increasing share of the labour force employed under non-formalized conditions (i.e. an increasing marginalization of workers from formal institutional relations) and the increasing but unrealized demand for qualified labour.

The Malaysian government has taken an unambiguous stand for the static flexibility 'model' by giving incentives to the exploitation of marginal labour reserves and import of foreign labour. The reverse strategy, as suggested by the Singaporean experience (see the chapter by David Drakakis-Smith in this volume), of deliberately raising wages to provide a strong incentive to upgrade to capital-intensive technologies, is not at all on the agenda.

From a regulationist point of view, traditionally concerned with underconsumption, technological development and (Keynesian) economic growth are, in the long term, hampered by these policies. Instead, state policies should give priority to a formalization of the labour market, thereby increasing local demand by raising real wages and stimulating dynamic technological development.

In the case of Malaysia, however, it seems less straightforward to pay attention to the issue of underconsumption. Due to the consumption pattern and the existing industrial structure dominated by TNCs, an increase in the wage level in the export industries would most likely add to the growing pressure on the balance of payments rather than provide a push for the virtuous circle of mutually increasing local demand and technology development as envisaged by this version of regulation theory.

The labour market is dominated by numerous non-formalized institutions. This is closely related to the fact that the present circumstances are marked by a booming demand for skilled and unskilled labour. This trend towards an extremely segmented labour market creates serious barriers for any effort to organize industrial wage labour and strengthen its bargaining power. Furthermore, the quest for flexible, modern craftsman-like skills-upgrading could be largely irrelevant as the technological path followed by the TNCs in the electronics industry – the all-important export industry – replaces qualified labour by semi-automated capital goods.

In this sense external defensive flexibility, including competitive wage setting, could be the most 'rational' form of the wage labour nexus to support the specific growth regime that has existed in Malaysia during the last two decades. In our view, the argument that only a wage labour nexus based on internal offensive flexibility (or dynamic flexibility in the 'local' terminology) will offer sustained growth in the long term is in essence biased towards the logic of Fordism as the conceptual centre of gravity for the regulation approach.

At the more specific level of Boyer's elaboration of growth regimes in outward-oriented Asian economies, this observation paves the way for answering the question as to why 'developmental states' actively pursue policies that – in Boyer's terms – maintain a 'bastard export-led growth' regime instead of policies aimed at transforming the economy into a 'virtuous export-led growth' regime. In the Malaysian case, we point to structural barriers to a rapid increase of productivity. But we also indicate the strength of the regulation approach as we look for links between economic growth patterns and institutional forms. Rather than identifying the attributes of an authoritarian regime, the regulation approach focuses on unravelling the intricate compromise between social forces and the institutional forms through which this is reproduced. Along these lines we offer an explanation of the apparently diffuse formation of a working class despite the presence of 'objective conditions' for the emergence of this social force brought about by the rapid industrialization.

At the centre of the mode of regulation in Malaysia lies a grand design for growth, distribution and modernization. Vision 2020 has become an ideological and political construction combining the aspirations of various ethnic and social groups. We conclude that the extent to which this vision has penetrated policy making and discourses in social institutions throughout the layers of Malaysian society is bringing the regime closer to a status of a 'hegemonic leadership'. Inspired by Gramsci's concept, Bocock suggests the following definition of this phenomenon: hegemony, in its most complete form, is defined as occurring when the intellectual, moral and philosophical leadership provided by the class or alliance of classes and class fractions which is ruling successfully achieves its objective of providing the fundamental outlook for the whole society (Bocock 1986: 63). We find that this quality of the state configuration in Malaysia must be fully understood in order to characterize the current phase in Malaysian capitalist industrial development.

Our methodological point is straightforward: research and development of the regulation theory must disentangle itself from wishful thinking about a reincarnation of Fordist regulation in a new industrializing part of the world. Social equality and parliamentary democracy are not implemented by well-intentioned economists but by struggling human beings; actually the theory reminds us that modes of regulation are shaped by social actors and not only by technocrats. Therefore, serious studies are needed of context-specific barriers to actions by social actors. These results could possibly explain why growing inequalities do not lead to the emergence of classes, i.e. class consciousness and forms of organization, and why new modes of regulation are established and transformed by institutional forms different from those in operation under Fordism.

Accordingly, we stress that the effects of particular labour institutions on economic development can only be evaluated through empirical studies. At present, there is a salient contradiction in the regulationist discourse between the general statements concerning the effects of different wage–labour nexuses and the quest for empirical sensitivity in the theoretical and methodological development.

1.   This selection limits the discussion by excluding a number of internal theoretical ambiguities within the regulation approach. We do not find this internal discussion irrelevant or sophistic but the purpose of this chapter is rather to discuss some of the more basic theoretical and methodological problems in relation to the application of regulation theory (in the above narrow sense) to the Malaysian industrialization process.

2.   The regulation approach has primarily been applied to studies of the capitalist development process in the industrialized countries and has focused on the establishment of Fordism in the USA during the middle decades of the twentieth century and the transfer of this regime of accumulation to Western Europe, notably to France and the Federal Republic of Germany in the initial decades after the Second World War. The Fordist regime of accumulation is based on mass production of standardized goods for price-competitive mass-markets. Semi-skilled labour makes up the major segment of the labour force. The corresponding mode of regulation is based on state-centred macro-economic regulation, public welfare programmes, and standardization of conditions in the labour market, notably the link between wage increases and manufacturing productivity increases through collective bargaining processes in the labour market. (See Harvey (1989) and Walker (1989) for reviews of the debate on Fordism and post-Fordism.)

3.   In addition five growth regimes for contemporary developing countries are outlined. The set of regimes is based partly on Lipietz's work (see above) with two further regimes added, a pre-industrial and a rentier, that – at a very general level – somehow fit the empirical realities of the majority of sub-Saharan African countries and major oil-exporting countries, respectively. Again, nothing new is revealed by this categorization and the reader is yet again confused by a new set of variables that constitute the basic elements of the growth regimes (Boyer 1994: 70–1).

4.   Part of the data presented was collected as part of a field research in Penang, Malaysia (July–August 1994). We are grateful to the Danish Council for Development Research for Financial Support.

Bibliography

Ahmad Sarji Abdul Hamid (ed.) (1993) *Malaysia's Vision 2020. Understanding the Concept, Implications and Challenges*, Pelanduk Publications, Petaling Jaya

Amsden, A. H. (1990) 'Third world industrialization: "Global Fordism" or a new model?' *New Left Review* **182** (July/August 1990)

Amsden, A. H. (1994) 'Why Isn't the Whole World Experimenting with the East Asia Model to Develop?' *World Development* **22**(1)

Bocock, R. (1986) *Hegemony,* Ellis Horwood Ltd, Chichester; Tavistock Publications, London

Bowie, A. (1991) *Crossing the industrial divide: State, society and the politics of economic transformation in Malaysia,* Columbia University Press, New York

Boyer, R. (1994) 'Do labour institutions matter for economic development? A "regulation" approach for the OECD and Latin America, with an extension to Asia', in G. Rodgers (ed.) *Workers, institutions and economic growth in Asia,* International Institute for Labour Studies, Geneva

Brookfield, H. (ed.) (1994) *Transformation with industrialisation in Peninsular Malaysia,* Oxford University Press, Oxford

Fold, N. (1993) *Linking Agriculture and Industry in Developing Countries: A Study of the Vegetable Oil Industry in Malaysia and Zimbabwe,* Geographica Hafniensia A3, Institute of Geography, University of Copenhagen

Harvey, D. (1989) *The Condition of Postmodernity,* Basil Blackwell, Oxford

Human Resources Development Council (1994) *Annual Labour Market Survey 1994.* Federation of Malaysian Manufacturers Northern Branch. Manpower Monitoring Committee, Penang

Jessop, B. (1990) 'Regulation theories in retrospect and prospect'. *Economy and Society* **19**: 2

Jesudason, J. V. (1989) *Ethnicity and the economy. The state, Chinese business and multinationals in Malaysia,* Oxford University Press, Singapore

Jomo, K. S. (1986) *A Question of Class: Capital, the State, and the Uneven Development in Malaya,* Oxford University Press, Singapore

Jomo, K. S. (ed.) (1993) *Industrialising Malaysia. Policy, Performance, Prospects.* Routledge, London

Kahn, J. S. (1994) 'New class contradictions between the urban and rural contexts in Southeast Asia'. Paper presented at the *Annual Conference of The Nordic Association for Southeast Asian Studies (NASEAS), 'Emerging classes and growing inequalities in Southeast Asia',* 23–25 September 1994, Tylstrup, Denmark

Kotz, D. M. (1990) 'A comparative analysis of the theory of regulation and the social structure of accumulation theory', *Science & Society* **54**(1)

Lim, L. Y. C. and **Pang Eng Fong** (1991) *Foreign Direct Investment and Industrialization in Malaysia, Singapore, Taiwan and Thailand,* OECD, Paris

Lipietz, A. (1986) 'New Tendencies in the International Division of

Labour: Regimes of Accumulation and Modes of Regulation', in: A. J. Scott and M. Storper (eds) *Production, Work, Territory: The Geographical Anatomy of Industrial Capitalism*. Allen & Unwin, Boston

Ministry of Finance (1994) *Economic Report 1994/95*, Kuala Lumpur

Ministry of Finance (1995) *Economic Report 1995/96*, Kuala Lumpur

Pillai, P. (1994) *Industrial training in Malaysia. Challenge and response*, Institute of Strategic and International Studies, Kuala Lumpur

Rasiah, Rajah (1994) 'Flexible production systems and local machine-tool subcontracting: electronics components transnationals in Malaysia', *Cambridge Journal of Economics* **18**

Rasiah, Rajah (1995) 'Labour and industrialization in Malaysia', *Journal of Contemporary Asia* **25**(1)

Standing, G. (1991) Structural adjustment and labour flexibility in Malaysian manufacturing: Some post-NEP dilemmas', in Lee Kiong Hock and Shyamala Nagaraj (eds) *The Malaysian Economy beyond 1990. International and domestic perspectives*, Persatuan Ekonomi Malaysia, Kuala Lumpur

Walker, R. A. (1989) 'Regulation, Flexible Specialization and the Forces of Production in Capitalist Development'. Paper for the *Cardiff Symposium on Regulation, Innovation and Spatial Development*, 13–15 September 1989, University of Wales

Wangel, A. (1991) *Areas of conflict and patterns of conflict resolution in industrial relations in Malaysia*. Working Paper No. 10, Centre for East and Southeast Asian Studies, University of Göteborg

Wangel, A. (1993) 'Technology transfer Japanese style – Direct investments in Malaysia'. Paper presented at the *7th General Conference of the European Association of Development Research and Training Institutes (EADI)*, 15–18 September 1993, Berlin

Yamashita, S. (1991) 'Japan's role as regional technological integrator in the Pacific Rim'. Paper presented at the conference on *'The Emerging Technological Trajectory of the Pacific Rim'*, 4–6 October 1991, Tufts University, Medford, Mass.

CHAPTER 8

The labour market in Vietnam: between state incorporation and autonomy

Irene Noerlund

Introduction

With the rapid industrialization in a number of countries in East and Southeast Asia, wage labour increases along with the proletarization of the relative surplus of agricultural labour. In many parts of the developing world, these changes do not usually lead to full or almost full proletarization compared with what happened earlier in Western countries. Small-scale production at the family level often absorbs a considerable amount of labour, which cannot be considered wage labour. Other people find employment in different kinds of informal production or services. The development differs from region to region in the world.

In Asia, one of the most dynamic economic regions, proletarization is more widespread because industry is expanding more rapidly than in other parts of the developing world. The labour market has, however, developed in a segmented way, with both formal and informal labour, and this has been an obstacle to the organization of labour and the establishment of trade unions. The question has been dealt with in many studies in relation to the first generation NICs (Newly Industrializing Countries): South Korea, Taiwan, Hong Kong and Singapore (Deyo 1989; Amsden 1989; Lauridsen 1992). In the case of the second-generation NICs or NICs-to-be like Malaysia, Thailand and Indonesia, the labour market has also been studied extensively from various perspectives (Limqueco et al. 1989; Lie and Lund 1994). In the third generation of rapidly developing economies, including China and Vietnam, the industrialization of the 1980s and 1990s has been studied quite thoroughly (Noerlund et al. 1995; Fforde and de Vylder 1988; Fforde and de Vylder 1995; Ljunggren 1993; Australia–Vietnam Research Project 1995), but in these cases the labour market and the organization of the labour force in trade unions have received less attention. As such, China and Vietnam differ from the other parts of Asia, because of their previously centrally planned economies and early attempts to build an industrial base. However, industrialization has never developed substantially in Vietnam; even less than in China. Since the start of reforms in the two countries from the end of the 1970s, experiments to develop 'market socialism' have started, with the purpose of reconciling the planned economies with the increasing importance of the market. From a classical Marxist perspective, the concept of market socialism appears to be a contradiction. In reality, what we are seeing is probably closer to a market economy with fairly strong state intervention. However, ideas of the planned economy will still have an effect for some time at least to come, and will influence the public sector of the economy.

The basic transformation of centrally planned economies, which Janos Kornai calls 'shortage economy' (Kornai 1982), into a market economy implies that the operation of the industrial sector is changed from what he calls the 'soft budget constraint' towards a hard budget. The soft budget implies that the industrial enterprises did not operate on the basis of profitability, as the state would usually supply the needed raw materials to a factory and take back the final product. The economic calculation at the factory consisted of a division of a specified percentage into various funds: wage fund, social fund and production fund. However, there was no price mechanism to determine the prices either of raw materials and other inputs into the production or of the final products on the basis of market realities; this made it very difficult to calculate whether an enterprise was profitable or not. What counted more was the importance and usefulness of certain products, which was a priority set by the state; factories with a high priority would receive the capital needed whether they were profitable or not. Moreover, wages were stipulated by law and only differed to a limited degree between five levels, depending on age and merit. The change into a hard budget calculation meant that market prices had to develop and that the enterprises would have to calculate their budgets on the basis of profit (or lack of profit, which in principle means that the enterprises have to close down). This had an impact on the internal organization of the factories, where productivity and surplus labour became an acute problem, with changes needed in order for the company to survive.

When the socialist regimes in the Soviet Union and Eastern Europe collapsed at the end of the 1980s, political reform dominated the situation simultaneously with the quick transformation into market economies. In the socialist countries of Asia, mainly China and Vietnam, the strategy has, on the other hand, sought to maintain the political regimes unchanged while introducing quite radical economic market-oriented reforms. In the case of Vietnam, the breakdown of the USSR and European socialist economies had a direct impact on the speed of reforms and external reorientation towards the market economies of Asia and the West. The strategy of 'market socialism' has so far been successful from an economic point of view. However, it is of interest to analyse the implications for the industrial work force and how it might react to the new conditions.

In the various countries of Asia, it is not possible to point to just one labour market model. In each country industrialization took place within the frame of a combination of the economic environment and state policy. In the broad perspective, the process three generations of NICs seem to run through involves a number of stages with respect to types of industry, starting with labour-intensive production in textiles and foodstuffs, then expanding into garments, leather and electronics. There seems also to be a trend towards a transformation from import substitution towards export-oriented production. Other elements such as the labour market, labour movement and its organization differ substantially from one country to another, however, and are closely linked to historical circumstances.

The purpose of this chapter is to present the transformation of the labour market in Vietnam, which is not only a third-generation industrializing country, but also one which is changing from a centrally-planned into a market-based economy. This has implications both for the industrial structures and the labour relations, and affects the workers' movement which used to be controlled by the party/state. The increasing importance of private capital, foreign capital, and the independence of enterprises and workers are bound to become highly influential factors in the socio-economic system.

Since the beginning of the economic reforms in Vietnam in the early 1980s, quite fundamental changes have taken place in the industrial sector, which was previously totally dominated by state-owned enterprises (SOEs). The enterprise reforms were, to a higher degree than those in agriculture, initiated from above, whereas reforms in the rural sector started in a spontaneous way at the grass-roots level and were subsequently approved by the central level. In the case of enterprises, compromises had to be found in the negotiations between the enterprise managers and the central level. The main change was that the enterprises were allowed to operate more freely after the fulfilment of the state-determined plan. The response from the workers was immediate and positive, since for the first time these innovations gave them an opportunity to achieve a higher income: piece-rate payment or bonuses related to the production result were concrete incentives for working harder for longer hours. Moreover, it was now possible to do extra work at home. The move towards a market economy started with the granting of permission to set up small-scale private enterprises at the family level, and the first links between the SOEs and market demand were established. But only when the enterprise reforms reached a more developed stage in 1987–88 and deeply influenced the existing system of labour relations, did it become meaningful to speak about the transformation or creation of a labour market. This stage was reached when life-time employment in the SOEs began to be discarded in order to make them profitable. New types of enterprise developed, because it was possible to set up larger private companies; foreign capital was invited from late 1987 and this resulted in the establishment of joint-ventures and 100 per cent foreign-owned companies.

Until the reforms of the early 1980s life-time industrial employment was guaranteed. The new generation of wage earners was usually recruited from among the existing workers' families as an important element in the existing strategy of maintaining working-class traditions, and wages were determined by seniority. Although the working class was first established by industrialization during the French colonial period and grew in the 1960s, it was limited in size, with no mobility between enterprises, and no wage competition. The new system was introduced in southern Vietnam in 1976 but it was difficult to speak of a labour *market* per se.

This chapter argues that, with the industrial reforms in Vietnam, a dual labour market is presently coming into being, mainly because of the differing characteristics of state-employed and privately employed work forces. Moreover, structures known from longer-established market economies in other developing countries such as increasing unemployment and conflicts in the labour market are emerging. Labour legislation is now being established in order to protect the interests of the workers and to regulate labour conflicts. In effect a fundamental change has taken place in the way the working class is considered by the state. Before the reforms, the proletariat was seen as an important component of the workers' and peasants' state. From this viewpoint no conflict between the working class and the state was deemed possible. Now, since the early and mid-1990s, as an example of the new societal thinking in Vietnam, legislation acknowledges conflicting interests between workers and employers. Nevertheless, it is questionable how far these new ideas have penetrated the labour system of the SOEs.

Start of reforms in industrial enterprises

In the years following the Vietnam War and the reunification of Vietnam in 1975–76, the enterprise system of the Democratic Republic of Vietnam (DRV or North Vietnam) was extended to the SOEs in the south. The first economic reforms in Vietnam began after the Sixth Plenum of the Communist Party in 1979, when the economy was in total crisis. They were first carried out in agriculture, but from 1981, reforms also started in the SOEs; these influenced the employment pattern in this type of enterprise. The former system based on largely equal wages, albeit with a graduated system of seniority of five levels, was changed into remuneration based on piece rate and new types of bonus. Moreover those enterprises where raw materials and investments were supplied and running expenses met by the state and the final products returned to the state were given greater freedoms to make extra products and to sell them on the market (the so-called three-plan system). Besides, the wage system involving piece rates became much more important than the basic wages: workers and their family members could have all kinds of extra work at home, with the raw materials being delivered from the factories and the final products going to the factories. The factories were, for their part, allowed to sell extra products on the market and keep the profit. The way was opened for higher incomes, increased production of consumer goods and the start of a free market. In a situation where lack of raw materials, electricity and spare parts were daily obstacles, the new system resulted in much harder work and longer working hours, but also in increased production, productivity and incomes (Noerlund 1990). The labour market, still characterized by life-time employment, started to expand mainly in the family sector.

Only after the decision on reforms known as *Doi Moi* in December 1986 were steps taken to change the enterprise system in a more fundamental way. In November 1987, Decree 217 gave the directors of SOEs substantial power to take decisions about the administration and management of the enterprises. The purpose was that the 'state-subsidies-bureaucratic-system' should be abolished in favour of a 'business-cost-accounting-system'. What Kornai called the soft budget constraint became harder, by allowing the profitable enterprises to develop more quickly and the deficit-hit enterprises to go bankrupt.

The immediate results were that the system of subsidized food and consumer goods, which was the old way of securing the basic necessities of state-employees, was abolished. Wages were increased as a compensation for the losses of some of the privileges and in principle regulated according to a price index based on the cost of rice (inflation was skyrocketing in 1987 and 1988). Basic wages still relied on a seniority system, but they were of secondary importance compared with piece rates and bonuses. In most factories, especially textiles and garments enterprises, the system of extra work disappeared or diminished in order to improve the quality of the products, and the workers were compensated through higher wages, still on a piece rate basis. In other factories where piecework was more difficult to apply individually, a collective piece rate was introduced, based on the over-fulfilment of the planned targets. However, people could still do extra work on their own account, i.e. get hold of raw materials and sell them on the free market, now as a private activity.

After the Vietnamese dong was made convertible in 1989, some of the large enterprises (with a turnover of several million US dollars) were allowed to have direct contact with foreign customers. Around 1992–93, the large state trading corporations, which had profited greatly from the subsidies system and their monopoly of trade, were reduced to the same level as other enterprises. Now the way was opened for enterprises to keep most of the profit they generated.

By 1995, the SOEs did not in principle have more obvious privileges than private companies. They had to take commercial loans whereas they had previously been able to borrow from the state at low interest. Even this privilege is apparently limited now. But as long as the SOEs are the backbone of the state economic policy, they will still be able to have access to various advantages, for instance tax exemption. In addition, the larger enterprises are not allowed to close down if they are considered by the state to be strategic. The market socialism today in Vietnam has to be understood as an economy basically operating on the principle of the market, but with strong intervention by the government to support the state sector, even if this does not appear in the public policy. This has happened to such an extent that the state sector, after a restructuring in 1989–90 and a rapid expansion of the private sector in the period 1987–91, is growing faster than the private sector (Table 8.1a).

The frame for understanding the economic reforms has to relate the change of priorities through compromises between the state and various economic and institutional interests manifested through the political system.[1] The pro-reform interests were supported by external pressures: one was the changes in Eastern Europe and the conflict with China, which forced a liberalization of production to overcome acute shortages; and the second was advice from institutions like the World Bank and the IMF later in the 1980s. However, the dynamics behind reforms basically come from internal forces. Among these are the pressures from directors and owners to be able to manage the economy of their enterprises more independently and from workers anxious to gain higher incomes and access to more consumer goods.

Unemployment: creating a labour market – the case of the SOEs

Piece rate systems were only introduced into the SOEs in the early 1980s and were widely accepted by the workers as an improvement on the former system which allowed very little scope for increase in income. However, it has to be borne in mind that the period 1982–94 was dominated by an almost constant growth in the industrial sector with only a few years as exceptions; usually, in periods of general expansion reforms are easier to carry out (Vu Tuan Anh 1995).

Only after the 1987 decision to introduce the cost-accounting system did the reform reach a more fundamental level, and it became clear that a number of SOEs had to be closed down because of their lack of profitability. The result of this was the necessity to sack a number of state-employed workers. This was most unwelcome to the political system in Vietnam, and because of the social commitment of the socialist government it took several years before these measures were carried out in practice. Over the years the government had proclaimed that it was important to secure the economic and social life of workers and state-employees and had put this into practice in the SOEs and in the cooperative system (handicraft cooperatives). To sack workers would break down the well-established principle of security in employment. In the SOEs a family-based employment system existed, where children were employed in the same factories as their parents. With the reforms at the end of the 1980s, life-time employment began to be abolished. The until then unknown concept of unemployment suddenly made its appearance.

Industrial production and labour in Vietnam

The statistical data on the Vietnamese economy is still not very reliable and sometimes figures are contradictory. To avoid misunderstandings, the period from 1985 to 1995 will be presented and analysed in some detail in order to develop the basis of the analysis of the economic changes and related labour situation.

We will look more closely at the changes in labour in the industrial sector, supplemented by the number of enterprises and small-scale production in which the workers are not considered workers as such, but still belong to the industrial sector. Finally the output of the various industrial sectors will be taken into account.

Table 8.1a Labour force in industry (not including construction) 1985–94 (000 persons)

	1985	1990	1991	1992	1993	1994
Total population	59,872	66,233	67,774	69,405	71,025	72,510
Total labour	26,020	30,294	30,974	31,819	32,716	–
Industrial labour	2,800	3,392	3,394	3,470	3,521	–
Total employment in state sector	3,859	3,421	3,143	2,975	2,923	2,933
Labour in state industry	870	808	704	686	687	699
Labour in non-state industry	1,930	2,584	2,690	2,764	2,834	–

Table 8.1b Labour force in industry (not including construction) 1985–94 (per cent)

	1985	1990	1991	1992	1993	1994
Industrial labour as % of total labour	10.8	11.2	11.0	10.84	10.76	–
Labour in state industry as % of total employed in state sector	22.6	23.6	22.4	23.1	23.5	23.8
Labour in state industry as % of industrial labour	31.1	23.8	20.8	19.9	19.5	–

Sources for Tables 8.1a and 8.1b: Dang Duc Dam, *Vietnam's Economy 1986–1995,* The Gioi Publishers, Hanoi 1995, Tables 1.1, 1.2, 1.4; Ministry of Labour, *Statistical Yearbook (Nien Giam Thong Ke),* 1993, National Political Publishing House, Hanoi 1994, Table 1.5.3.

The structural breakdown of the labour force in Vietnam gives a percentage of industrial workers at around 11 per cent of the total working population (Table 8.1b). It could be argued that the figure for the labour force is low, for instance the World Bank estimated it to be 36 million in 1993 (*World Development Report 1995*). A higher figure would lower the industrial employment rate, which might be reasonable in a comparative perspective. In other Southeast Asian countries the rate differs: Philippines wage labour/non-wage labour in industry 12.3/3.7; Indonesia 10.0/5.8; South Korea 30.5/5.1; Thailand 8.7/3.2; Malaysia 19/3.6. In China, however, the rate is estimated to be 4.0/14.0.[2] These figures can obviously only be taken as indicators. They are calculated differently and include both wage and non-wage labour. Wage labour seems especially low for China and is obviously calculated differently from figures for Vietnam. The non-wage industrial labour is also large, which can be seen from other indicators (Table 8.2).

It is an important feature of the industrial structure in Vietnam that state-employed workers only constitute about 20–30 per cent of the in-

Table 8.2 Number and structure of enterprises 1985–92

	1985	1990	1991	1992
Number of state industrial enterprises	3,050	2,762	2,599	2,268
Cooperatives	35,629	13,086	8,829	5,723
Private enterprises	902	770	959	1,114
Private household	–	376,900	446,771	368,000

Sources: Dang Duc Dam: *Vietnam's Economy 1986–1995,* The Gioi Publishers, Hanoi 1995,
Table 11.8.

dustrial labour total, whereas the state share in employment is much
higher in China and up to 70 per cent in Russia (Adams 1994). However,
the share of state-employed workers fell sharply from 1985 to 1991 when
it levelled out at around 20 per cent. The share of industrial state labour
out of the total state-employed work force did not change much in the
period 1985–94, though it increased slightly (Table 8.1b). This points to
the fact that the whole state sector was undergoing a process of change
and reduction of labour in this period, i.e. not only industry. In actual
numbers the labour force in state industry was reduced by 184,000 from
1985 to 1992. This includes the fluctuation which in the period 1985–90
saw an increase of state labour up to 1988, after which it started to fall, so
the reduction took place within about 4–5 years (*Industrial Data* 1989–93,
1994). However, the very important development was that the total
amount of industrial labour increased in the period 1985–90, so that the
non-state industry was able to make up for some of the lost jobs in the
state sector. This calculation does not, however, include the increasing
labour force in general during this period, and quite a lot of workers had
to be absorbed within agriculture. In fact with new arrivals competing for
industrial jobs, increased unemployment occurred in the industrial sector,
especially at the end of the 1980s and in the early 1990s.

The second indicator of industrial development is the number and type
of enterprises (Table 8.2). The number of enterprises changed considerably in the period 1985–92. In the state sector, the number was reduced by
about 25 per cent. This is less than the reduction in the labour force, and
in fact a number of enterprises were amalgamated, with the labour force
being reorganized in a more productive way. The number of private units
seems to have been stable, but this hides the fact that private enterprises
were very small in 1985, having only started to operate freely around 1988
when the number was quite limited. The cooperative sector is clearly in
the process of being eliminated, having experienced a reduction from
35,629 enterprises in 1985 to 5,723 in 1992. On the other hand, many were
transformed into private units or household enterprises. Unfortunately
household production was not calculated before 1988 but by that year it
already consisted of 318,000 units. It is not safe to conclude that household production simply took over from cooperative production, but it certainly increased from the early and mid-1980s, when it was encouraged
by state policy. It is important to note that this small-scale industrial production is of considerable scope and indicates that industrial labour in the
non-wage sector represents a substantial force, although it is difficult to

calculate exactly. It might be even larger than the official statistics reveal as some small-scale production may not be registered. Moreover, the few figures presented in Table 8.2 also reveal that this sector is characterized by considerable fluctuation in size. Although household production has been encouraged by the government, it has certainly not received economic support from the state. It is usually left to market forces, because official policy still gives priority to the state sector.

The total picture cannot be understood from the figures in Tables 8.1 and 8.2 only. From 1988, the private sector was vigorously revived in Vietnam; however, after a few years the state attempted to strengthen the state industrial sector, both through the restructuring of the enterprises and through various more or less hidden measures.

The gross output index gives a fairly good impression of the substantial growth-rates in state industry and the private sector, whereas the cooperative sector is playing an increasingly smaller role (Table 8.3). The private sector has the highest growth, but that should be seen in the context of the contributions to the GDP by the state and the private sectors, whereby industry plays the dominant role, contrary to the larger picture (Tables 8.4 and 8.5). Even if the total non-state industrial sector is weakened by a shrinking cooperative sector, the difference between the state

Table 8.3 Gross output of various industrial sectors 1985–94 (index 1985 = 100, at constant 1989 prices)

	Total	State	Non-state of which:	Cooperative	Private
1985	100	100	100	100	100
1986	106	106	106	114	94
1987	117	116	118	122	112
1988	134	134	133	123	123
1989	129	131	127	78	166
1990	133	139	120	63	183
1991	147	155	136	37	232
1992	172	187	149	25	294
1993	193	212	162	25	324
1994	219	241	183	–	–

Sources: Dang Duc Dam: *Vietnam's Economy 1986–1995*, The Gioi Publishers, Hanoi 1995, calculated from Table 11.1.

Table 8.4 Structure of GDP in state and non-state sectors (constant 1989 prices, per cent)

	1990	1991	1992	1993	1994
State	32.8	33.7	34.7	35.7	36.9
Non-state	67.2	66.3	65.3	64.3	63.1
Total	100.0	100.0	100.0	100.0	100.0

Source: Dang Duc Dam: *Vietnam's Economy 1986–1995*, The Gioi Publishers, Hanoi 1995, Table 1.20.

Table 8.5 Structure of GDP in the industrial sector divided into state and non-state (per cent)

	1990	1991	1992	1993	1994
State	66.0	66.6	68.7	69.3	69.3
Non-state	34.0	33.4	31.3	30.7	30.7
Total	100.0	100.0	100.0	100.0	100.0

Source: Dang Duc Dam: *Vietnam's Economy 1986–1995,* The Gioi Publishers, Hanoi 1995, Table 1.20.

sector in industry and its contribution to GDP in general is striking. It shows that the state considers the publicly owned industrial sector as a key to the economy and therefore also gives it high political priority. Moreover, the attempt by the state to strengthen and expand the state sector seems to have been successful, in spite of the dynamic new private sector.

The conclusion may seem contradictory. The state sector is, on the one hand, numerically not very large either in numbers of enterprises or in labour employed, but on the other hand state enterprises are generally large ones. The restructuring attempt has been successful and the state sector is continuing to play a key role in the industrial strategy as far as output is concerned. The industrial development, which at a first glance today seems to be dominated by foreign investments and small-scale production at the household level, is still to a large extent dominated by the SOEs. This may be the reasoning behind the two-pronged policy of the state, which encourages foreign investments and small-scale production whilst trying to maintain the importance of the state sector. One of the ways to maintain political leverage, which cannot be seen from the statistical data, is to combine SOEs with joint-ventures, thereby giving the state sector a certain influence in the joint-ventures with foreign companies.

Development of a segmented labour market

One of the obvious reasons why the reduction of employment in the state sector happened so quietly is connected to the economic expansion in general. This was mainly due to the growth of jobs in the private sector and household production. Joint-ventures with foreign companies created about 60,000 jobs[3], while the household economy contributed considerably to job creation. As in other countries, it is very difficult to know the exact size of the informal sector or household economy.

However, official unemployment increased from a low level in the crucial years from 1988 to around 1992 to an estimated rate of more than 10 per cent in the cities in 1994. A new type of labour market in Vietnam was in principle established as a result of the new policy allowing the SOEs to fire workers and the need for labour in the new type of enterprises. Theoretically, the hiring and sacking of workers is now based on principles of cost and benefit. The drive for the establishment of the labour market comes from the new economic structure which transforms labour into a commodity. The new labour market is segmented with respect to wages, social security and labour organization. Another important reason for the workers to accept the loss of jobs in the SOEs is their special type of enterprise system, which works on the basis of consensus between the various partners involved at the factory level.

Before the reforms, the employment system was characterized, on one hand, by labourers in life-time employment in the SOEs and cooperatives. The other type of employment consisted of the self-employed, which although it barely existed on paper did so in reality, and followed the fluctuations of the 'quasi-existing market'. Even if self-employment did exist, because it existed in a grey zone between legality and illegality the form was limited because of restrictions of the commercial markets. With the establishment of a labour market, a new division in employment developed. Firstly, there are the workers of the SOEs, who are no longer lifetime employed, but contract workers. Secondly, there are the workers in larger joint-ventures with foreign capital which are regulated separately. The third type of enterprise is in the private sector (including joint-stock companies), small domestic units which are either self-employed, developed small family enterprises or even larger ones. The main new feature of the labour market is competition for jobs, for salaries and in some circumstances also for workers.

The SOEs have reduced their labour force, but in general they try to maintain as many employees as possible.[4] Many SOEs have reduced the incomes of new workers after restructuring the labour force, which was necessary in order to restructure state industries. When new workers are hired, the traditional employment pattern of employing young workers who will continue in the factories on a long-term basis is often repeated, because the enterprise-system has not completely adopted real market mechanisms. However, a new trend has made its appearance in the most advanced factories: they now seek to attract skilled or semi-skilled outside workers, whereas previously labour would mainly be trained on the shopfloor. Criteria like a good education and skill are becoming more important, but at this point mainly found in southern Vietnam. In several factories advertisement and competition between candidates are new ways of recruiting labour.

The reduction of employees in SOEs is mainly due to the number of workers who have left voluntarily or to the closing down of companies. The reason why they left voluntarily has to be seen in the context of the labour system in the SOEs which has not changed very much. Even if the legislation acknowledges the different interests of workers and employers, the institutions at the workplace are still intertwined in a corporative way with disputes, to a very large degree, finding solutions within the institutional framework of the enterprise. In this respect industrial relations differ fundamentally in the SOEs and the foreign-owned enterprises. It could be argued that class consciousness is very different compared to the Western market-oriented employment system as Vietnamese workers show greater solidarity with the company and are much less ready to struggle for their individual interests. Even compared to China, labour conflicts seem to be fairly few (Kaye 1994) as Vietnam's work force appears to be more docile.

Researchers from an institute in the Labour Ministry confirmed that most of the workers left the SOEs voluntarily. The explanation was that they could get jobs in the private sector, which expanded quickly from the end of the 1980s, and that the wages were higher in the private companies. Visits to various parts of the country confirmed that the wages in private firms were often, but certainly not always, better. The private enterprises have only absorbed a small proportion of the workers; small-scale production and trade (informal sector) are probably the largest source of employment. The income levels in the various sectors are, moreover, diverse. Some of the private enterprises pay better wages, whereas others do not (for instance private garment factories in Nam Dinh paid less than SOEs). If business in the informal sector or small-scale firms is good, the income is often the same as in the SOEs or even higher, sometimes much higher. Before the reforms there was both greater prestige and a higher level of privileges connected with employment in SOEs than nowadays. In general, some of the privileges like cheap food and other basic needs have disappeared. Workers received other types of compensation, however, as late as 1993, the workers in Nam Dinh could take over the living accomodation provided by a large textile factory on condition that they themselves would now be responsible for the maintenance of the buildings. In the southern countryside this took place even in 1995. But benefits like housing are of course, also an incentive to keep the job, thus reducing the mobility of the employee. Some of the SOEs pay their workers well when the business is doing well. There is, however, no longer a unified wage system among the SOEs, and in some cases the workers get no wage for several months if the company is in trouble.

Differences in income and working conditions

From visits in the country, I found the typical wage level in the SOEs in mid-1994 in Nam Dinh to be 300,000–400,000 dong (US$30–40) monthly (prices are lower in Nam Dinh than in the larger cities), in Hanoi around 400,000–500,000 dong and in Ho Chi Minh City 500,000–600,000 dong (US$50–60). In the Ho Chi Minh City area wages are in general higher than in the north, but living costs are also higher. The salaries averaged around US$50 a month in the textile and garment factories. In one of the SOEs in the south, a shoe factory, workers had an average income as high as US$60 a month in 1994 but in this case it was related to work in a dangerous environment with quick-drying poisonous glues and chemicals applied in the processing of leather products. The other factory which paid salaries of around US$60 was a smaller private garment company in Ho Chi Minh City. In this case it was compensation for not receiving any social security. The pay depended totally on the work the workers had delivered. The lowest salary in the south was found in a Korean–Vietnamese joint venture which gave the workers a wage of around US$40.

The incomes are in principle based on a combination of basic wage, bonus and a piece rate-based pay. In the 1980s, piece rate and bonus became very important; however a new wage system in the 1990s decided that bonuses in the SOEs cannot constitute more than 50 per cent of the regular salary.[5] The piece rate system still seems to be an important element of the wage, and it is considered to be the best system, for instance by the Ho Chi Minh City Confederation of Labour, because the skill of each worker is valued and it encourages workers to raise the standard of labour and do away with laziness. In the opinion of the Confederation, even the workers prefer this system and it is promoted by the city union.[6] In the private factories and some joint-ventures the piecework system is the only way of calculating wages, whereas in the SOEs, the system of grade based on seniority combined with various incentives is still in existence. However, the importance of piece rate systems differs from one kind of activity to another depending on the type of work. For instance, the piece rate system is much easier to apply on an individual basis in a garment factory than in a sugar factory.

The main and very important difference in working conditions in the various types of labour-intensive factories is the length of the working day. In the largest factories with three shifts, the working day would usually be 8 hours, which is also stipulated in the Labour Code of 1994. In case of overtime work the payment should increase to 150 per cent of the normal wage. But especially in garment factories the working day is usually much longer, as high as 10–12 hours. In the written contract, which is now obligatory between employer and employee, in a private joint-venture it was not stipulated how many hours the workers were supposed to work; according to the director this depended on the number and time-frame of orders. In a private garment factory, the workers were, for instance, sewing on a Sunday just like any other day. The owners explained that it was the workers' own choice if they wanted to work in their factory. As long as labour is plentiful the bargaining power of the workers is not high. In the SOEs, social security is much better than in the private factories: sick pay, maternity leave (although it is being cut from 6 to 4 months) and holidays are still available, even if some of the facilities from the subsidized time have been reduced, such as housing and child care. In any case, new measures are often undertaken to compensate for some of these losses.

Joint-ventures with foreign capital are more closely monitored by the state and the party than private domestic enterprises and they have to implement regulations more strictly than the other types of enterprise. The South Korean director in a joint-venture said that he tried to avoid extra labour above the prescribed 8 hours, because he would have to pay the workers 150 per cent in overtime and that would mean a loss for the company. But of course, if there were problems in finishing an order in time it was necessary to work overtime, he added.

All in all the working conditions, including social security, are generally better in the SOEs than in other types of enterprises, with joint-ventures coming next and the private factories last of all. However, some of the private factories have the advantage of paying higher wages to 'compensate' and even if there is no social security the relation between the employers and employees will often take on the character of 'family relationship'. When a child is born, for instance, the owner will give some extra support to the workers. On the basis of field-work and without general statistical data, it is hard to know if the higher wage level in private companies is a general phenomenon, as it seems to differ from the north to the south and from company to company. The observations might indicate that the competition between firms to acquire labour is more widespread in the south than in the north and that the labour market is starting to function more on a market basis in the private sector in southern Vietnam.

It seems reasonable to talk about a segmented labour market: on one hand the SOEs with more regulations, a higher level of social security and a lower level of labour mobility; and on the other hand a labour market for the privately-owned enterprises in which wages might be higher, but where social security is at a much lower level, quite often with family-like relations between owner and employees and a higher mobility of labour. The third major group consists of the joint-ventures with foreign capital, which are more strictly regulated than the nationally owned enterprises, but carry with them cultures of employment and labour relations from other countries. In general they are less interested in the local rules, traditions and influences of the surroundings, though it depends very much from which country the foreign capital originates. Reactions from the workers have also to be seen in this perspective.

Whither the trade unions?

Ever since the creation of the Democratic Republic of Vietnam – which was proclaimed after the August Revolution of 1945 – the labour unions have been an integrated part of the political system based on the alliance of workers and peasants under the leadership of the Communist Party. The role of the trade unions does not seem to have changed very much over the years, even after the reforms started. However, new trends can be found. The trade unions are considered a mass organization with the main purpose of supporting workers in the factories and in private life. In doing so the function of the unions is to contribute to the increase of industrial production, which is considered essential to improving the economic situation of the country and in this way also the livelihood of the workers. Some labour structures and the role of unions are, however, changing under the impact of the development of the market system.

The wage systems brought about by the reforms based on bonus, piecework and extra work in the early 1980s had already introduced new elements of inequality, which had to be discussed among the workers as this could be a potential source of conflict. After Decree 217 in 1987, which gave the directors of factories more autonomy, workers' influence also increased at the factory level. The appointment of directors by the central level had to be approved by the workers' congress; but this was more of a formality than an expression of real influence. The awareness of the new rights was probably limited at first. When the factories started to dismiss workers in the SOEs, it suddenly concerned the workers directly and discussions took place among the workers to decide criteria for dismissal.

This is probably one of the first instances where workers were directly involved in a decision of such vital importance to their future as life-time employment was coming to an end. Of course, the fact that the workers were involved in this decision-making process also helped the management board to avoid conflict.

At the congress of the Vietnam Workers' Union in 1988, the name was changed to the Vietnam General Confederation of Labour (VGCL). Nguyen Van Linh, the Secretary General of the Communist Party, made a speech at the congress which urged the unions to act more forcefully and independently of party and management. He mentioned the interesting fact that leading trade union cadres did not have to be party members (*Vietnam Courier* 1989). The change of name indicated that it was not only a union for workers but for all kinds of labour. It should also be mentioned that the new elected chairman, Nguyen Van Tu, has been able to give a higher public profile to the VGCL. After the fundamental economic reforms in 1987–89, the attempt to institutionalize a society based on laws started and a long process of establishing basic labour legislation started. In 1990, a number of decisions were promulgated: a law on Trade Unions, a decree introducing contract labour, and regulations concerning labour in foreign-owned enterprises were decided by the Council of Ministers, followed by provisions from the Ministry of Labour (known as Decree 233). In 1992, the new general constitution of the state was promulgated and, directly related to the trade unions, a new constitution of the VGCL was passed in 1993. For quite a long period, the content of a Labour Code was discussed, and it was finally passed in the National Assembly in June 1994.

Before the passing of the Labour Code, a debate took place about whether or not trade unions should be established in the joint-venture sector. The argument against it was based on the fear of scaring investors. At the National Assembly meeting in 1993 it was decided that trade unions should be established.[7] The increasing number of cases of bad treatment of workers by foreign employers probably had an impact on this decision. Thirty drafts were in fact discussed before the final version was widely published and debated in 1993–94. This draft was also considered at the factory level, although only on the management board (usually consisting of the director, the deputy director and the accountant) and in the board of the trade union. The workers would not discuss the labour code at shop floor level until after it was passed. The factories were outlining collective agreements in mid-1994 for acceptance by the directors and the trade unions even when only the draft of the Labour Code was known. This was the first written agreement to regulate the different positions of the trade union and the management board. However, even in early 1996 only a fraction of the companies had signed collective agreements.[8]

The Labour Code of 1994 was directed mainly at the state-owned and private sectors. Decree 233 concerning labour in foreign-owned enterprises (1990) had to be adjusted in accordance with the Labour Code; this took place during 1995. In most respects the Labour Code now defines relations between labour and employer, including conflict resolution. The most heated debate focused on the level of minimum wages, the right to strike, and some social measures like maternity leave which was shortened from 6 to 3 months. The discussion on minimum wages did not finish before the final passing of the Code in June 1994. So far, the monthly minimum basic wage is 120,000 dong (US$12) in nationally-owned companies and for foreign-owned enterprises it was set in 1992 at US$30 in Hanoi and US$35 in Ho Chi Minh City. At first, in 1990, it was set at US$50 but was lowered in 1992 as a result of pressure from foreign companies. In 1990, US$50 was considered a high wage, whereas in 1994 it was about average in many larger well-run companies. In 1996, it is an average or even below average salary. Prices rose by 36 per cent between 1992 and 1995.[9] A spokesman from the Labour Ministry said, shortly before the Labour Code was passed, that the basic wage for an unskilled worker would increase to US$50 when the code was finalized, but that proved too optimistic a statement. Even in 1995 the former level was maintained. The debate continued, and the VGCL had in early 1996 obtained the government's approval for a proposal to set the minimum wages in joint ventures at US$50.[10]

A representative from the Ministry of Labour had difficulty in understanding why some 100 per cent foreign-owned enterprises should be opposed to the organization of unions since they are not meant to initiate conflicts but rather to facilitate communication between employers and workers. It is well-known from other countries that local officials are against trade unions because it may scare off foreign investors. Officials all over the country were in favour of the establishment of trade unions,[11] which would help to institutionalize the dialogue between the director and the workers. The role of the unions is, moreover, to benefit the enterprises' business. The theory is that the more successful the enterprise the better the conditions for the workers; the unions still have a role in supporting workers with social problems and in cultural life. New items on the agenda are training and labour protection. The trade union cadres visit the workers at home if they are ill, there are special funds for funerals, and for children who lose their parents, etc. Even when there are disputes in the families the trade union can play a role. This is probably a deeply-rooted function in the existing organization-oriented employment system in the SOEs with its close relation between factory work and family-life. With the introduction of the Labour Code an important limitation of the trade unions took place because they previously used to administer the social security system of the factories. Now the security system has become the responsibility of an insurance company. The system has only been in operation since late 1995 and experience of it is still limited. Factory-level unions, however, complain that they could use the money more effectively than the insurance company, which will keep any profit for itself and not for the benefit of the workers.

The Labour Code took effect in January 1995. The optimistic goal was to establish trade unions or preliminary workers' councils in all factories within 6 months, and trade union representation on the management boards. The financial base of the trade unions has become more independent from the state which means in practice that there is less money available. The unions receive no support from the state except basic facilities like accommodation and means of transport. The income of enterprise union officials consists of 2 per cent of the enterprise's wage fund and the workers have to pay another 1 per cent of their salaries. Of this amount the first 1 per cent of the wage fund is contributed to higher-level unions. However, the unions can undertake other types of income-generating activities.

The role of the unions to date has to be seen within the framework of the party. The political structure in the enterprises is the so-called '*bo tu*' or 'the four interests': party, government (management), trade union and youth union. Today the administration has a more important role to play than before, but it is primarily related to the business aspects of the enterprise. The party decides the general political line of development and the administration carries out the decisions. After the directors gained more power in 1987 with Decree 217, a dilemma might have arisen, if the directors and the core party members were not the same people, because the secretary of the party traditionally ranked higher than the director. This dilemma can be avoided by combining the position of manager and party-secretary and it happens quite often. In the SOEs the management board usually consists of 3–7 members, almost all being party members. The trade union leader will often be a high-ranking member of the party but that need no longer be the case.[12] The trade union is represented on the management board and management are also members of the trade union like all employees in the factory with the exception of foreigners.

The collective agreement that had to be set up in many factories from 1994 is an agreement between the director and the trade union leader. It means that the rights of workers have been set down on paper in a way which has not happened before. This might lead to problems for the directors if, for instance, wages cannot be paid or not paid on time, or if the workers have to work overtime with no extra payment. The question is, however, whether the traditional institutions in the SOE will intervene before a conflict develops.

The unions play an important role at the provincial level. 'The first principle in the work of the unions is to defend the interest of the workers', said the vice-chairman of Nam Ha's provincial trade unions (Red River Delta in the north); other functions are ideological work, organization of movements and help with social problems. One of the main purposes is, moreover, to create more jobs in order to reduce unemployment. The trade unions themselves have been reorganized and the number of employees at provincial level was reduced in Nam Ha from 70 to 30.

The interesting new aspect is what happens in the joint-ventures and the private factories where the party cannot claim to have a legitimate right to be the leading partner and does not have the right to be represented on the management board. The solution is probably easiest in joint-ventures with an SOE as the partner. When a trade union is established in the joint-venture, the party will play its role through a mother company or directly in the factory through the union. In one enterprise a party cell was established directly in the foreign-owned factory, and in another case the party members of the joint-venture enterprise were members of the party cell in the mother SOE. The evidence shows that a uniform system does not exist. The factory which had a trade union put a question to the Vietnam General Confederation of Labour about the role of trade unions in joint-ventures; this is a most relevant question, because suddenly the trade union has a different role to play in the structures of a foreign company without 'the four interests' and the interwoven institutional set-up of the SOEs.

In joint-ventures not related to an SOE or in the case of privately owned enterprises, the trade unions are only about to be established. In some cases they are set up because of pressure from the workers; in most cases they are not established, either because of resistance from the owners or lack of interest on the part of the workers. The Labour Code as mentioned promulgated that trade unions should be established within 6 months of a factory being set up. Twenty per cent of the joint-ventures had established trade unions in mid-1994 (*Vietnam Economic Times* 1994). By 1996, the percentage of unionized joint-ventures increased somewhat to about 20–25 per cent.[13] In the newly set up Export Processing Zone of Tan Thuan near Ho Chi Minh City, the trade unions have had greater success, with unions established in 14 out of 34 enterprises, i.e. 41 per cent.[14] This is evidently more a result of state policy to establish unions than a broad workers' movement.

The provincial trade union is responsible for setting up local trade unions. Although the Constitution of the VGCL of 1993 makes it possible for grass-roots units to be set up in enterprises and to be approved afterwards, the small-scale and medium-scale private sector is probably the most difficult sector to unionize; there are also differences between the northern and southern parts of the country. The union system is generally more accepted in the north because of its longer tradition there. The private owners do not feel obliged to pay the 2 per cent of the wage fund, however. Nevertheless, it was found that even in a family-owned rural enterprise in the Red River Delta with about 200 workers, where the family did not belong to the Communist Party and did not want to join it by any means, a positive attitude towards setting up a trade union prevailed.

From the fragmentary evidence available to us, it appears that the level of unionization of private factories is higher in the rural area in Nam Ha province, where 17 out of the 200 private enterprises had set up trade unions by July 1994. In a suburban area around Hanoi only 3 out of 60–70 private enterprises had unions in late 1995.[15] Figures from the south indicate a low rate. In Dong Nai province, bordering Ho Chi Minh City, out of the 700-odd mainly small private companies none had set up trade unions. It had only happened in some of the joint-ventures.[16] Visits to Ho Chi Minh City also indicate a lack of interest in establishing unions among factory owners in the small-scale private sector.

The question of wages and working conditions is probably more important for the workers in the private sector than the question of whether or not they have a trade union. The prospects for unionizing the industrial enterprises depend to a large extent on the campaigns carried out by the existing trade unions. The Labour Code is an important instrument in this respect. After it was promulgated, an apparently large-scale information system was established to inform workers about the content of the long and complicated text. It is one of the main obligations of the new trade union members to recruit further members, for instance in the EPZs (Economic Processing Zones).

All in all, the picture of the role of the trade unions is blurred. The basic structures of society are changing and a new type of union has begun to emerge in the non-state sector, which attempts to defend labour interests. The most important unions are still in the SOEs, however. The role of labour is becoming more defined and regulated in general as the number of unions expands. The structure of the unions is changing, with the establishment of new professional branch unions in more fields. Nevertheless, the enterprise unions and the provincial unions are still the dominant feature of the union structure, and the enterprise system in the SOEs still links the unions closely to the party. It probably depends on the unions in the individual enterprises to use the new measures which are promulgated and to shoulder their new responsibilities. Internally, the unions are still closely linked with the party at the factory level. For the workers, it is probably most important that they are still considered to be the core of the party; many are members of the party[17] and the policy is to maintain comparatively good living conditions for (state) workers. Institutionally, the structures of the enterprise-systems of the SOEs are being maintained, but the new policies on labour might lead to more changes.

Labour conflicts

The labour regime is, in spite of everything, changing more slowly than the economy in general. We shall briefly look at one of the areas which shows that workers are on the move: the labour conflicts that have taken place within the last couple of years. Conflicts in the SOEs have traditionally found solutions within the enterprise system and the strikes which have taken place cannot be taken as representing conflicts in the labour market in general. But it is a new element of Vietnamese labour relations, and therefore interesting as an indicator of change. Labour conflicts were illegal until February 1994, when the decision was made to allow strikes on condition that 15 days' notice was given. In the Labour Code it has now been confirmed that if 50 per cent of the workers want to strike the action is legal.

The number of conflicts recorded in official sources differs. Some authorities give a very low number. The VGCL stated that in the period from 1989 to mid-1994, 100 conflicts took place.[18] ILSSA gave the number of 90 from 1987 to mid-1994. The last source stated that the number of conflicts has decreased in recent years: 70 cases took place between 1990 and 1992; less than 20 cases were recorded in 1993–94.[19] In the first half of 1995 – after the Labour Code came into force – a source quotes the number of 20 (AFP 1995). In Ho Chi Minh City, after the promulgation of the Labour Code, the number of strikes for 1995 was recorded at 21,[20] and for the early months of 1996, 9 strikes were recorded in southern Vietnam before the lunar New Year (Tet) by the newspaper of the VGCL, 'Labour' (*Lao Dong* 1996). The trend is not clear; although the number of labour actions recorded is not high, remaining more or less at a constant level, possibly a little lower in Ho Chi Minh City. However, the introduction of the Labour Code has not been able to stop conflicts and most of them still take place as wild-cat strikes, because the regulations of the Labour Code are very slow and complicated (Noerlund 1995). However, one of the important reasons for strife was due to lack of implementation of the Labour Code.[21]

The conflicts took place not only in foreign-owned enterprises and joint-ventures, but also in SOEs, in public offices, and in the newly established professional unions (like street sweepers and taxi drivers) mainly in Ho Chi Minh City, Hanoi and Da Nang (only one was heard of in Nam Dinh). However, the conflicts have not been very large (one of the largest included around 600 workers) and not very long-term, lasting at most 2 weeks and usually only a few days. The largest conflicts have so far developed in enterprises without trade unions; in joint-ventures they were of larger scale than in the SOEs. Three main reasons are outlined as reasons for conflicts in joint-ventures up to mid-1994[22]:

1. Related to the wages. Some of the joint-ventures only paid US$20 (the law stipulated US$35). In South Korean- and Hong Kong-owned enterprises they agreed to pay US$35 after a few days' strike.
2. In joint-ventures the working conditions are sometimes very bad and the workers are forced to work up to 12 hours per day without overtime payment.
3. There are cases of bad treatment of workers and punishments like beating-up workers, letting them stand in the sun, etc. Local governments have intervened in some cases.

Conflicts in the SOEs are generally smaller and the reasons for conflict are most commonly related to unequal distribution of the incomes of workers in the factories.[23] Some of the conflicts are known from other sources and the information does not contradict that from ILSSA. Probably the largest conflict took place in April 1994 in a cashew-nut processing plant in Song Be province in the south and involved 800 workers. The newspapers wrote about this conflict, but in general labour conflicts are not an important subject in the newspapers.

Most conflicts have happened in the south, and in quite a number of cases they took place in enterprises owned by South Korean and Taiwanese capital. There seems to be a clash of employment and management patterns, where the East Asian systems introduce much harsher labour discipline than that found in Western enterprises and SOEs. However, one of the most important sources of conflict is the lack of a new-year bonus in the foreign-owned companies. This is a long-established tradition in the SOEs. Enterprises with capital from other countries might introduce their systems also. Four hundred out of 1000-odd joint-ventures are situated in the south around Ho Chi Minh City and this will probably influence labour mobility, as wages increase more quickly here than in other parts of the country.

The question is why the Labour Code accepted strikes as a means for the workers to raise their demands when in fact very few people are really eager to use this measure, including many workers. The new role of the trade unions and the establishment of collective agreements in the factories are ways of turning conflict into a formalized dialogue between workers and employers. The workers who actually achieved results from their illegal strikes were, however, those who were supposed to have higher incomes but had not always done so.

In a joint-venture dominated by foreign capital independent of a SOE, the situation is different. A company, for instance, 70 per cent owned by South Korean capital and headed by a Korean director, had several labour conflicts. According to the legislation on trade unions from 1990, the workers can establish unions, and this was suggested strongly by the workers (especially one who later became head of the trade union). The Korean director was against the idea from the very beginning, because 'in Korea the workers want to demonstrate every day'. He had to accept the Vietnamese legislation after several strikes; and the relations between the director and the workers improved considerably after the establishment of the union because he suddenly had a counterpart with whom to discuss the problems. The conflicts in this company had been among the largest in the south, with 650 people staging a strike.[24] The problems were related both to the level of wages, which were below US$35, and cultural problems between the Koreans and the Vietnamese.[25] The two large strikes took place in August 1992 and February 1993 in this company[26] and the trade union was set up in 1993. However, since the establishment of the union no big problems have developed. This interview took place in the presence of both the South Korean director, the Vietnamese deputy-director and the head of the trade union. They all agreed that the situation worked much better between the employer and the employees, and they had clearly found a *modus vivendi*. At the same time it was obvious that the employer had to follow the official rules, and in this case, the trade union had a considerable influence.[27]

Another example of a factory which encountered labour conflicts was a textile and garment company, a combined SOE, which had established three satellite joint-ventures with South Korean and Taiwanese capital. In this case the conflict started because one of the South Korean supervisors beat up a worker when the workers were on their way home. They were meant to do some extra work but did not understand what the supervisor said. The trade union chairwoman described the situation in considerable detail. She saw the situation as a misunderstanding resulting from the lack of a common language. The workers reacted with a short strike and asked the union to take action. The union demanded that the supervisor be sent back to South Korea. However, the board of directors could not agree with this measure, and suggested that an apology was enough. Shortly afterwards a general meeting was called and the supervisor had to give a public apology, but he did not lose his job and stayed in Vietnam.

Although there have been conflicts in other joint-ventures with, for instance, German management, it seem as if problems have more often developed in enterprises with South Korean and Taiwanese capital. But it should also be borne in mind that to a larger extent than capital from other countries South Korean capital is invested in labour-intensive production and Taiwan is the largest investor. The problems appear to concern stricter labour discipline, low wages and cultural clashes. From the party's side this is also regarded as capitalist exploitation which should be dealt with.

It seems that the situation differs from one company to another with regard to types of conflict and especially union strength. Clearly the idea behind the Labour Code was to reduce the number of labour conflicts but it may also help the workers, who know their rights much better than before. Moreover, there might be companies which are not interested in trade unions and might put pressure on local authorities, especially in the south.

A smaller private company in Ho Chi Minh City had quite a different approach to labour legislation and labour conflicts. First of all it was not interested in setting up a trade union. There had not been problems with the workers, and the young female owner said mildly that, if problems developed, the workers were free to leave; nothing forced them to stay. The Labour Code was not regarded as a big problem, because, as she said, so many new laws were promulgated but not implemented. The workers worked long hours, but they were compensated by higher wages, and, the owner added, she and her husband were working even longer hours. She found they had good relations with the workers and generally had no problem in keeping them. This is an example of a new trend of development in the private labour market, especially in the south of Vietnam and in the largest cities.

The trade unions are generally better established in the north and have been so for a longer period. However, the unions in the south have functions more similar to those of Western trade unions and investments are larger there. This combination is probably the reason for the more widespread and spontaneous labour conflicts which have appeared there.

Conclusions

When the secretary general of the Communist Party, Do Muoi, said in August 1994 that the state protects both owners' and employees' legitimate interests (*Vietnam News* 1994), it is difficult to know how much it will actually support the workers in cases of labour conflicts. Strikes are not regarded as an appropriate weapon, because of the danger that foreign capital will be less attracted to come to Vietnam, but also – and not least – because the conflicts are expected to find solutions within the factory if possible.

Labour conflicts are nevertheless the most visible indication that the labour market has changed and even more than that: the whole labour regime, including state, enterprises and industrial relations, is taking a new shape. From the available sources, the trend is that most of the conflicts have taken place in the private enterprises, mainly joint-ventures, even though conflicts have also been observed in other types of factory. The joint-ventures might in one respect be an area where both the state, the unions and the workers, can to a certain extent agree to conflict, because this type of factory is, according to the socialist rhetoric, representative of 'foreign capital' and 'foreign exploitation'. Therefore, it is interesting to note that it is not only in this type of enterprise that conflicts have taken place. The economic reforms have brought changes in the enterprise system which have led to the creation of a new type of labour market. However, the transformation is slowed down by the institutional and cultural setting. Nevertheless, the reforms of the economy – opening up to foreign capital and the creation of a labour market, combined with the increasing importance of labour and trade unions – have led to more initiatives by the workers. Moreover, the conflicts are more numerous in the south than in the north, which indicates that foreign investment and private economy are more developed in the south. This should be related to the much longer tradition of the socialist labour regime in the north than in the south.

It is debatable, however, whether the Labour Code will promote a more active defence of workers' rights or not. The purpose of the Code is, after all, an attempt to avoid conflict in the labour market. On the other hand, the whole process of establishing labour legislation has involved the workers to a higher degree than before and their rights are set down on paper. This gives the workers and the trade unions more scope to raise their demands, especially in the foreign-owned enterprises but also in the SOEs. The small-scale private sector will probably take a long time to unionize, if at all, and it is less likely that labour conflicts will take place in this sector because of the family-oriented relations of production.

The unions have a new role to play, especially in the joint-ventures, and there are cases where the unions have gained a certain degree of autonomy. It is still too early to make a judgement on their role in private enterprise. The significance of ideology should be taken into consideration; the many years of talk about securing the lives and interests of the workers as the fundamental basis of society does not disappear from one day to the next but the role of the unions also depends on action by the workers and the trade unions (and the party) in the various factories. In some factories the workers are completely docile and the enterprise system has not changed fundamentally from the former system of interlinked interests of labour, management and party. In others, the workers use the opportunities given by the new conditions and the new legislation and in a number of cases they have reacted spontaneously to protest against labour conditions. The segmented labour market also contains areas where labour relations are not based on class relations, and neither trade unions nor labour conflicts are going to have much influence there.

The government and party are probably not sincere about the use of strikes as a weapon. The legislation is too complicated to make the option of striking an easy one. Officials and factory directors agree that they did not find strikes an appropriate means of expressing dissatisfaction. One of the reasons why the Labour Code has been passed now in such a detailed form, with its 198 articles and more than 100 pages, is probably related to the general push to modernize the legislation and to be in line with international trends – and demands. The International Labour Organization's legislation was an important source of inspiration in the formulation of the Labour Code, and Vietnam has officially signed a considerable number of ILO clauses.

The chairman of the Vietnam General Confederation of Labour, who is also a member of the Central Committee of the Communist Party and a member of Parliament, sees a new role for the trade unions. But he also recognizes that the entire trade union system has been in a turmoil in recent years (Nguyen Van Tu 1993; JPRS-SEA-94-002 1994). He finds that class consciousness has been weakened over a longer period because skilled workers have been allowed to earn a living by crop cultivation, animal husbandry and even trading. The trade unions have also been passive because of the subsidies mechanism (Nguyen Van Tu 1993; JPRS-SEA-94-002 1994). In an interview published by Doan Ket he said: 'If workers wait for the trade unions to be organised for them, or when trade unions are organised only after there is collective struggles in the establishments outside the state sector, it is too late' (Dai Doan Ket 1993; JPRS-SEA-94-001 1993). He suggests that the trade unions should set up unemployment and welfare funds, help each other when workers take leave, are ill or unemployed; set up funds to develop the family economy, and for creative activities like sports clubs; improve living conditions; and protect the legitimate rights of workers. Only then will there be a movement in which the workers will believe, and they will be committed to the trade union organization (Dai Doan Ket 1993; JPRS-SEA-94-001 1993).

This is a vision of a society not very different from that of social democracy a century ago in Europe, based on a growing labour movement. Because of state policy, trade unions will now be established in Vietnam to a larger extent than in most other countries of the region. There are many conflicting interests which will play a role in the coming period in the state, the unions and among the workers. The question is whether a state which claims to be socialist and which is facing rising demands from the workers will be able to achieve a result different from what has been seen in a number of capitalist East and Southeast Asian countries? Will it result in the rise of an increasingly market-oriented labour system and a possible diminishing role for the party? In Vietnam the party still has a grip on the political system, which has both an impact on the policy towards labour, which is favoured in the SOEs, and on the maintenance of the enterprise system to a considerable degree in the SOEs in which the corporative interests are institutionally linked together. This combination of an expanding market economy and state regulation leads to a new type of segmented labour force, which is not only based on different wages and social security or labour in different types of branch. The main difference is the corporate type of decision which takes place in the SOEs and the strength of the workers in this process. The situation of the enterprise depends on its side on actual performance, the importance of production for the national economy, and relations between the enterprise and the central level. In this respect, political relations become as important as economic performance. The joint-ventures and private enterprises will to a much higher degree than the SOEs depend on market forces, with labour relations based on cost and benefits. This is a type of segmentation different from that of other countries with plural political regimes.

However, it is argued that in spite of the maintenance of many institutional structures in the enterprise system, there are changes not only in the economic patterns, but also in the trade unions and the legal system. These indicate an increasing openness to making the institution of the market also influence the SOEs, as in the abolishment of life-time employment. The question is therefore whether the party-state is going to take reform even further, including privatization of the SOEs. This was on the agenda for discussion at the party congress in June 1996. In a more fully privatized economy the enterprise system of the SOEs will also have to be changed. The present system is still in a period of reform and pressures exist both to maintain the status quo and to bring about further changes. The workers in the SOEs would probably lose out if the system were to change, but that would abolish the special segmentation of the working class.

Placing the case of the reform of the Vietnamese economy in a regional perspective, it may be interesting to note that the only country with a similar labour regime is China. Research on the similarities and differences could contribute to a better understanding of transition to 'market-socialism' (Chan and Noerlund 1995).

1.   I am basically in agreement with the research carried out by Melanie Beresford, Adam Fforde and David Marr in the Australia–Vietnam Research Project. See for instance Beresford (1995). It is a matter of debate whether industry is fundamentally working on the principle of the market or whether it is still continuing on the basis of a centrally planned economy. Beresford is arguing for the latter.

2.   World Development Report (1995). Unfortunately Vietnam is not included in the World Bank's statistics.

3.   Information from Ministry of Labour, Institute of Labour Science and Social Affairs, 26 July 1994.

4.   Except for the workers who were doing a very bad job who were sacked first.

5.   Interview with the chief of section of labour and wages in the Ho Chi Minh City branch of the Labour Ministry, 29 February 1996.

6.   Interview with the director of the Ho Chi Minh City General Confederation of Labour, Mrs Hoang Thi Khanh, 5 March 1996.

7.   Communication from Prof. Mac Duong, head of the Institute of Social Sciences in Ho Chi Minh City, 27 February 1996.

8.   Interview with Labour Confederation by Schwarz (1996).

9.   Schwartz (1996). Statistical data show an increase of 38 per cent January 1992–January 1995. Dang Duc Dam (1995).

10.   Interview with the director of HCMC General Confederation of Labour, Mrs Hoang Thi Khanh who is also the vice-chairperson of the VGCL, 5 March 1996.

11.   This was in the period after it was decided in the National Assembly that trade unions had to be established.

12.   In the Bason Shipyard, an old enterprise established during the French colonial time, the chairman of the trade union was very knowledgeable about all aspects of the enterprise's performance, and he was not a party member. This was accepted after the trade union congress in 1988. In this case the trade union chairman is in a very special position, because he is also member of the executive board of the Vietnam General Confederation of Labour.

13.   20 per cent of joint-ventures should be organized, quoted from Tran Dinh Hoan, Minister of Labour, December 1995/January 1996, *Far Eastern Economic Review* (25 January 1996). Sai Gon Giai Phong quoted 25 per cent in February 1996.

14.   Interview with the president of the Labour Union of the EPZ, Ms Doan Thi Thu Ha, 29 February 1996.

15.   Communication from Flemming Carlsen, consultant to the General Workers' Unions in Denmark from a visit in December 1995.

16.   Communication from party member at provincial level, March 1996. This was a fact even if the owners of the small enterprises are virtually all members of the party. This indicates, however, that the party has an important role to play among the small-scale private enterprises in general. Maybe the large cities are the exception.

17.   16 per cent of the state workers are considered members of the party. Thanh Tuyen (1996).

18.   Interview with VGCL, 30 July 1994.

19.   In Ho Chi Minh City, however, 36 conflicts were registered in 1994. Labour Department, February 1996.

20.   Interview with the director of the HCMC General Confederation of Labour, Hoang Thi Khanh, 4 March 1996.

21.   According to the Labour Department, February 1996.

22.   Interview with the Institute of Labour Science and Social Affairs of the Ministry of Labour, 26 July 1994.

23.   ILSSA, interview 26 July 1994.

24.   Information from the General Federation of Trade Unions.

25.   Information from the General Confederation of Trade Unions in Hanoi, confirmed by the director.

26.   Information from the General Confederation of Trade Unions in Hanoi.

27.   A strike was staged at this factory in February 1996.

Bibliography

Adams, Nina S. (1994) *Workers' Education on Woman Workers and Child Labour*. Report from study trip to Hanoi 16/21 March 1994, for the ILO

AFP (1995) *Vietnam Labour Laws Fail to Halt Strikes*, Hanoi, 9 July

Amsden, Alice (1989) *South Korea and Late Industrialisation*, Oxford University Press, New York and Oxford

Australia–Vietnam Research Project (1995) *Researching the Vietnamese Economic Reforms: 1979–86*. Monograph Series #1, January, School of Economic and Financial Studies, Macquarie University, Sydney

Beresford, Melanie (1995) 'Interpretation of the Vietnamese Economic Reforms 1979–85', in Australia–Vietnam Research Project, *Researching the Vietnamese Economic Reforms: 1979–86*. Monograph Series #1, January, School of Economic and Financial Studies, Macquarie University, Sydney

Chan, Anita and **Irene Noerlund** (1995) Vietnamese and Chinese Labor Regimes: On the Road to Divergence, paper presented to the 2nd workshop *Transforming Asian Socialisms: China and Vietnam Compared*, Temporary China Centre, The Australian National University, August 1995. Forthcoming in a revised edition in *Transforming Asian Socialisms* (preliminary title)

Dai Doan Ket (1993) Hanoi Weekly 6–12 November

Dang Duc Dam (1995) *Vietnam's Economy 1986–95*, The Goi Publishers, Hanoi

Deyo, Frederic C. (1989) *Beneath the Miracle. Labor Subordination in the New Asian Industrialism*, University of California Press, Berkeley

Far Eastern Economic Review (1996) 25 January

Fforde, Adam and **Stefan de Vylder** (1988) *The Economic Transition in Vietnam*, SIDA, Stockholm

Fforde, Adam and **Stefan de Vylder** (1995) *From Plan to Market. The Transition in Vietnam 1979–94*, Westview Press, Boulder, Colorado

Industrial Data 1989–93 (1994) Statistical Publishing House, Hanoi

JPRS-SEA-94-001 (1994) 4 February

JPRS-SEA-94-002 (1994) 18 February

Kaye, Lincoln (1994) 'Labour Pains. Workers' Unrest Could Challenge the Party's Legitimacy', *Far Eastern Economic Review*, 16 June

Kornai, Janos (1982) *Growth, Shortage and Efficiency*, Basil Blackwell, Oxford

Lao Dong (1996) 'Cac vu tranh chap lao dong trong dip Tet', 3 March

Lauridsen, Laurids S. (1992) *Labour and Democracy in Taiwan? Continuity and Change of Labour Regimes and Political Regimes in Taiwan*, Research Report 87, Department of Geography, Roskilde University, Denmark

Lie, Merete and **Ragnhild Lund** (1994) *Renegotiating Local Values*, Curzon Press, London

Limqueco, Peter, Bruce McFarlane and **Jan Odhnoff** (1989) 'Labour and Industry in ASEAN', *Journal of Contemporary Asia Publishers*, Manila and Wollongong

Ljunggren, Borje (ed.) (1993) *The Challenge of Reform in Vietnam*, Harvard Studies in International Development, Harvard University Press

Nguyen Van Tu (1993) *Tap Chi Quoc Phong Toan Dan*, October

Noerlund, Irene (1990) 'Textile Production in Vietnam and the Philippines. The Development of Female Labour 1970–85', in Marja-Leena Heikkila-Horn and Jouko Sappanen (eds) *Southeast Asia: Contemporary Perspectives*, Finnish Association of East Asian Studies, Helsinki

Noerlund, Irene (1995) *Labour Laws and the New Labour Regimes. A Comparison of Vietnam with China*, paper presented at the Conference: Vietnam . . ., Centre for Pacific and Asian Studies, Stockholm, August 1995

Noerlund, Irene, Carolyn Gates and **Vu Cao Dam** (eds) (1995) *Vietnam in a Changing World*, Curzon Press, London

Schwarz, Adam (1996) 'Proletarian Blues', *Far Eastern Economic Review*, 15 January

Thanh Tuyen (1996) 'Cong doan voi cong tac xay dung dang (The trade unions support the building of the party)', *Lao Dong*, No. 29/95, 7 March

Vietnam Courier (1989) 'The Sixth Congress of the Vietnamese Trade Unions', No. 1

Vietnam Economic Times (1994) 'Striking a Balance'. Interview with Nguyen Van Tu, President of the General Confederation of Trade Unions, May

Vietnam News (1994) 'Communist Party Deliberates Industrialization', 2 August, No. 1079

Vu Tuan Anh (1995) 'Economic Policy Reforms', in Irene Norlund, Carolyn Gates and Vu Cao Dam (eds) *Vietnam in a Changing World*, Curzon Press, London

World Development Report (1995) *Workers in an Integrating World*, Table 25, Oxford University Press, New York

CHAPTER 9

Inconsistencies and inequities in Thai industrialization[1]

Somboon Siriprachai

Thailand has been proclaimed to be an NIC (newly industrializing country) by international organizations such as the World Bank (1993), but the country faces acute problems that are rarely discussed. Although economic growth has been impressive, the areas of human development and income distribution are deteroriating. In fact, this pattern of uneven development can invariably be seen as a trade-off between the agricultural and industrial sectors. For years, the rural poor have been trying to solve their present difficulties by persistent migration to seek work in Bangkok, principally in the so-called 'informal sectors'.

The severe effects and externalities of the industrialization process have been paid less attention. In fact, the production structure is characterized by the hallmarks of an agrarian society. Although manufactured exports have been growing rapidly, at a rate similar to those of the East Asian NICs in the 1960s and 1970s, population and employment structures are still overwhelmingly rural-based. The share of the agricultural labour force was 70 per cent in 1986, but only declined to 65 per cent in the 1990s. It suggests that a long-term deterioration of rural living standards is quite likely, particularly in terms of a low consumption rate.

The aim of this chapter is to analyse whether the economic policy reforms of the past two decades may have been too few and too late to solve the problem of rising unbalanced growth and to eradicate poverty. The analysis adopts a neoinstitutional economic framework to explore the role of the state, and to learn from economic history in Thailand. It points to major institutional changes that are most essential to bring about agricultural modernization and determine how far industrialization can proceed. The persistence of the squeeze on the agricultural sector will be a significant hindrance to continuing sustainable economic growth. Rapid increases in the productivity of agriculture will be of major importance for domestic demand to absorb manufactured goods. This sector should in turn be subsidized by the state to improve productivity and living standards. These conclusions are reached through a comparison with the experience of the NICs in economic policy reforms and the role of the state.

A conceptual framework

Thailand began its integration into the world economy in the middle of the 1880s and emerged as the major supplier of rice to the rest of the world, in particular the ex-colonies of Great Britain (Ingram 1971). Modern Thai economic growth, however, only began in the late 1950s. Most of the Thai leaders during this period showed serious interest in transforming the country from an agricultural to an industrial society. Yet, the Thai state might be considered inclined to being predatory to the extent that depletion of scarce resources has taken place. The predominant economic strategies returned little in terms of public goods. The consequences have been low productivity in agriculture, weak human capital development and poverty in the rural areas.

There are five reasons for considering the Thai state as predatory and not developmental:

First of all, the outcome of economic development in the last four decades suggests that the state was entirely controlled by the ruling elite: the military, vested-interest groups (mostly ethnic Chinese) and bureaucrats who seemed eager to protect their own interests without providing adequate welfare in terms of education, health care, provision of a pure water supply and adequate sanitation.

Second, the state seems to be unique in the sense that it has been rather weak in making any coherent policy effective. The most crucial policies (e.g. land reform and income distribution policy) have been implemented, but in a somewhat erratic manner which proved insufficient in the end. This is intertwined with the unbounded power of the bureaucracy, which might have led to arrogance, overconfidence and an anti-democratic attitude. The East Asian NICs (except Hong Kong) were very strong and effective in terms of the degree of relative autonomy from vested interests in decision-making. The Thai state has been relatively weaker as it has been unable to insulate itself from vested-interest groups. The assumption that the state is well-intentioned, well-informed and competent appears to be wrong. In reality, the state might be manipulated and incompetent as in the Thai case. Nevertheless, Thailand is fascinating in the sense that, since the 1930s, the armed forces have never held absolute power as in Indonesia, South Korea or Taiwan. The monarchy, being the absolute source of legitimacy, has been above politics and makes Thai political economy unique.

Third, the main failure emerges from the fact that the bureaucracy, which evolved from the Thai *Sakdina* (feudal) autocracy, has turned out to be a huge and clumsy monster (Riggs 1966; 1993). The rigidity of this hierarchical system has created a klepto-patrimonial instead of a meritocratic bureaucracy (Evans 1995). In fact, there has been no substantial change in the civil service since King Rama V reformed it. Power was exercised by the King in the past – today this is done by technocrats and bureaucrats. The merit system embedded in the bureaucracy has been gradually deteriorating to the system's disadvantage in that material incentives are rather low despite a very rapid transformation of the country's economic development (Christensen and Siamwalla 1993). The bureaucratic class has dropped in status and the ethics of the organization are weak. But the power structure is still entrenched due to the survival of authoritarian regimes. The legal authority for administrative measures and the operation of the bureaucracy as determined by the King in the old days have been divorced from public liability and accountability (Rosenberg 1958). This is prevalent in the Thai political structure, which has remained unchanged after 60 years of semi-democracy or the so-called 'soft-authoritarian regime' (Chamarik 1981; Samudavanija 1992). The military, the technocrats and their business (Sino–Thai) and academic allies have repeatedly been appointed to the senate in this period and are instrumental in maintaining a kleptocracy. Hence, discretionary power and arbitrariness have been almost untouched by administrative law. Although the rule of royal servants was broken after the people's coup in 1932, civil servants have taken over and manufactured obedience.

Fourth, it is worth noting that the Thai state has been unable to foster long-term entrepreneurial perspectives among indigenous private elites, thereby lacking the one crucial type of state activity mentioned in Gerschenkron's typology: the state must create the so-called 'state entrepreneurial actions to substitute for missing entrepreneurship'. In reality, transformation has somehow been impeded rather than promoted by the state, resulting in a relatively limited number of Thai entrepreneurs in commerce and industry. Moreover, commercialization in the countryside is extremely insignificant and dominated by ethnic Chinese middlemen who are seldom autonomous, but subservient to autocrats, aristocrats, and lately, the military.

Fifth, the political regime has been (soft) authoritarian despite the very short period of democracy in 1973–76 and in the 1990s. Agriculture has been heavily squeezed to lend support to the industrial sector. Productivity in the agricultural sector appears to have been stable albeit increasing at a very low level due to the resource endowment and land-abundancy until the late 1970s (Siriprachai 1995). The initial condition of rich natural resources made it possible for the Thai state apparatus to extract and dissipate rents when industrialization was initiated. In other words, Thai peasants could continue to exhaust natural resources for longer compared with resource-scarce countries like the NICs. Under different conditions, the agricultural sector could have contributed to economic growth without turning forest land into cultivated areas, if the state had generated rapid technological progress. The so-called 'green revolution' never had a chance to take root in the economy owing to the shortage of irrigation water. Existing dams, drainage and reservoirs are inadequate for success in developing and disseminating the new technologies.

Furthermore, the infant industries such as those producing fertilizers and agricultural machines, sold at high prices to the agricultural sector, are protected by high import tariffs. Thus, high economic growth was extensive rather than intensive. Besides, it has been concentrated in Bangkok, which is often cited as the city in the world with the highest level of primacy. This concentration of growth and industrialization in the metropolitan area has inevitably given rise to uneven development and inequality leading in turn to an income gap. The performance of industrialization suggests that intersectoral conflict and accumulated rural–urban antagonism will occur (Aeusrivongse 1993). The developmental states of the East Asian NICs cannot simply be emulated by Thailand, because the state and institutional settings are different in their historical and social contexts. How and why these factors matter will be analysed below.

The Thai state might have functioned relatively well in protecting individual rights, people and property and enforcing voluntarily negotiated private contracts, but it has failed to create a strong domestic market and eradicate poverty in the rural areas. The strategy of the developmental state is quite difficult to implement, largely due to the relatively backward agricultural economy. The state has been playing a contradictory role in not closing the productivity gap between agriculture and industry. Thus, a backward dualistic economy continues to exist. Although the state, since the late 1950s, has attempted to industrialize through import-substituting strategies, this seems unlikely to be successful because of the difficulty in adapting new technological skills and absorbing surplus labour.

This shortcoming is further constrained by an inflexible institutional factor, particularly in its key institutions: the lack of a hard state or relative state autonomy to threaten unproductive firms (Amsden 1989; Wade 1990). Moreover, it has been forgotten that the experience of the East Asian NICs lay not so much in foreign investment and multinational corporations, but in national capital and indigenous industrialists as well as in rapidly building up an educated populace. This entails fostering internal articulation and achieving technological autonomy. In fact, ISI (import-substitution industrialization) might not have failed in Thailand had it been accompanied by substantial improvements in agricultural productivity and income distribution that could in turn absorb domestic manufactured goods. Such a process has always been taken for granted by the Thai ruling elite, who were obsessed with the market mechanism as a means of linking the ISI industries to the relative backwardness of agriculture.

The point is that productivity in the agricultural sector has not been substantially upgraded as opposed to the industrial sector. In fact, agriculture was dominated by the vestiges of *Sakdina*, with cultivation techniques dating back to the previous century. In addition, the bureaucracy is not professionally competent enough to prevent the ruling elite and entrenched interest-groups from extracting and distributing unproductive rents. It is undeniable that a rent-seeking society is always closely linked to public policy and allocation of scarce resources. Thus, increased social injustice and income inequality not only remain unchanged, but have tended to become more acute and more widespread.

Since the middle of the 1970s, the Thai state has been changing its strategy to one of export-led industrialization, but without solving fundamental problems, i.e. the inadequate infrastructure, the lack of skilled labour and low secondary-school enrolment, the low productivity in the agricultural sector and impoverished rural inhabitants who are mostly farmers. Although the new development strategy seemed to keep economic growth high, mainly through the influx of direct foreign investment, the basic failure of the industrialization process, as mentioned above, still prevails today under the same institutional setting. In fact, the institutional arrangements appear to have made matters worse in so far as the rise of money politics and vote-buying have been encouraging corrupt politicians. In addition, the relative backwardness of agriculture has presented an unintended opportunity not only for the urban bourgeoisie and ex-technocrats, but also for local capitalists and influential persons who ran for Parliament. Thus, electoral politics is believed to have paved the way for them to legitimize their power under money politics. However, to understand the political economy of Thai industrialization, we must turn back to the ISI (import-substitution industrialization) policy in the late 1950s.

State promotion of industrialization

Under the Sarit government, industrial promotion policy was renewed by the Promotion of Industrial Investment Act of 1959. The Board of Investment (BOI) was set up in the same year. This period saw the second rudimentary institutional change since the Chakri Kings had undertaken political and economic reforms in the mid-nineteen century. The multiple exchange rate regime and large-scale state enterprises were supposed to come to an end. The former was replaced by export taxes and a fixed exchange rate, while the latter became concerned with infrastructure development. The role of the Thai state was also believed to be that of a social guardian or a benevolent dictator and to provide a stable investment environment for the private sector (Chaloemtiarnana 1979).

In this period, the state removed the 'invisible foot' and put back the 'invisible hand' of the market mechanism into the field of production, permitting private enterprises to develop. The result of the Sarit regime's use of the Revolutionary Party had a major impact on the power structure of modern Thai political economy, namely, the monarchy, the bureaucracy and the military. In politics, it essentially interrupted and weakened Parliament and political parties. From 1932 to 1963, the military was seldom accountable to civil society. Thus, Thailand has hardly had the social and economic bases for a democratic development. The middle class, which was predominantly Chinese and therefore treated as alien, allied itself to the ruling elites. Very few parliaments completed their terms (4 years). They often had to adapt their role to new circumstances (Meesook et al. 1988).

However, the legacy of Sarit to Thai economic history was the concern created by the increased role of technocrats.[2] There is little doubt that, to a certain degree, the stable macroeconomic performance, notably economic growth and low inflation rate, has partly been brought about by these dwindling, able and honest technocrats (Thanapornpun 1990). While technocrats give priority to planning and economic engineering, the bureaucrats who approve and grant rights are the patrimonial officials. This is a sharp distinction in modern Thai bureaucracy.

Turning to the structure of production, ISI could not be regarded as successful in the 1950s to the 1960s. Heavy protection was implemented through an ad hoc sectoral policy in the form of high tariffs, import quotas and other quantitative restrictions, especially import surcharges. However, it is misleading to take for granted that the high tariff barriers were made available as a tool for industrial policy protection under a strategic industrial policy regime. On the contrary, it has from time to time been deployed as a source of government revenues by the Ministry of Finance (Patamasiriwat 1993). ISI should have substituted the imported goods through increased production within Thailand to meet the demand of the markets, but resources were instead transferred to the preferred industries by the BOI (as promotion certificates) through the provision of relatively cheap machinery and intermediates, mostly capital and inputs from abroad. Furthermore, the high tariffs imposed on finished products under ISI inevitably led to the import of more capital and intermediate products for assembly only.

Thailand has neither been guided by industrial sectoral planning nor by any industrial targeting strategy as in the East Asian countries. This is indicated by the fact that:

> The Thai State does not control the markets for credit and foreign exchange, thus depriving policy makers of perhaps the key tools for conducting industrial targeting. Furthermore, there has been little co-ordination or coherence in the use of existing industrial policy instruments, tariffs, investment promotions, capacity controls and local content regulations. (Christensen and Siamwalla 1994: 18)

The BOI, being one of the main authorities for implementing industrial policy, did not establish a coherent import-substitution policy *per se*; nor did it build up a clear-cut framework to oblige targeted firms to meet its requirements. Punishment has never been the case in the Thai context. The one explicit principle of the BOI is to encourage foreign companies to invest, no matter the kind of industry. Later, some criteria such as labour absorption were included. The obvious evidence is that no performance standards and follow-up have been imposed on the recipients of promotion certificates.

There were four prerequisites for making ISI successful in East Asia: the strength of the state, competent bureaucrats, independent technology learning capacity and increased productivity in agriculture prior to EOI (export-oriented industrialization). This implies that income distribution must be more equitable to consume manufactured goods in exchange. Such a process properly raises rather than lowers standards of well-being. Consequently, the incentives in industrial policy pursued by the BOI, namely the granting of tax holidays (8 years), exemption from import duties on machinery, components and raw materials and the imposition of bans and surcharges on competing imports enabled the BOI to seek economic rents rather than enhancing productivity through adapting modern knowledge.

Bribery started to play a significant role when industrialists, both foreign and local, applied for promotion certificates. However, the BOI is by no means a rent-seeker (Meesook et al. 1988). Not surprisingly, most incentives (taxes) offered by the BOI to firms are opportunity costs to all Thais. Scarce resources distributed to promoted firms will be increased twofold unless these firms succeed in adapting and mastering new technological skills. Linkages became weak as an expanding manufacturing sector was not able to generate activity throughout the entire economy. Hence, the ISI industries are at best foreign enclaves as long as agriculture is left behind.

It can be concluded that the BOI has played an active role in the allocation of the nation's production resources by encouraging specific sectors and discouraging others through the tax system.[3] However, the BOI is clearly biased against agriculture and this is contrary to the experience of the NICs. Thus the package of incentives seemed not to stem from a coherent strategic industrial policy (Siamwalla 1993).

It has been argued that the East Asian NICs also provided plenty of incentives, but why is the performance of the manufacturing sectors in Thailand, by comparison, inadequate in terms of labour absorption and technological content? A closer look at the contextual and institutional factors seems to be essential for an understanding of the course of the Thai industrialization process. As a matter of fact, a high degree of discretion and selectivity in the granting of incentives for a wide range of objectives without any accountability to civil society, accompanied by little monitoring or follow-up of promoted enterprises, made it different from the experience of the East Asian NICs (Christensen and Siamwalla 1994). The implication is that technology is more than tools, plants or machines: it is very complex and embedded in key institutions. It also becomes clear that the Thai state has not created an indigenous technological capacity to support technological transfers from foreign firms.

As pointed out by Marzouk (1972), the encouragement of capital-intensive (or labour-saving) techniques with a low contribution to employment led to inefficiencies in the allocation of scarce capital resources. To some extent ISI rarely enabled the economy to avoid a constantly increasing trade imbalance and deteriorating balance-of-payments. In particular, the labour-absorbing capacity of this strategy in the context of the economy appears to have been rather weak. This has adversely affected progress in agricultural modernization and has erratically resulted in swelling numbers of urban poor.

This is entirely contrary to the industrial policy adopted in Japan, South Korea and Taiwan. One explanation is intimately related to the relative autonomy of the state and the role of a strong national development ideology in these countries (Siriprachai 1993). The three East Asian NICs underwent lengthy periods of agricultural improvement before ISI. They also succeeded in establishing relatively large growing domestic markets for manufactured goods, linking industry to higher productive and dynamic agricultural sectors. The most essential contributor is the state's capability to build up a large literate population with limited resources in a short time.

Export promotion and the legacy of ISI

ISI industrialization in Thailand left many problems unresolved. By the early 1970s, the leaders had begun to turn away from ISI and adopt EOI. The question is, why did the state decide to make this turnabout?

In 1963, the Bank of Thailand had warned of the many problems of ISI. However, with economic growth during the first two decades of its post-World War II history, the Thai economy was quite satisfactory, but not outstanding: GDP growth averaged 5.2 per cent in the 1950s, rising to 7.4 per cent during the 1960–72 period. As the World Bank mission stated, ISI began to face problems of excess capacity as the market became saturated in the late 1960s.

ISI, in reality, became the preferred strategy in Thailand not because of the rational arguments recommended by the World Bank, but rather because of expeditious policy actions to meet balance of payment crises. Needless to say, there was a common interest in ISI on the part of the bureaucratic authoritarian state, urban manufacturing entrepreneurs and transnational corporations. But how much and how long ISI should continue to depend on protection was cause for debate.

The infant industry argument was implicitly used. It was widely known that ISI in Thailand was not targeted according to systematic economic criteria as in the East Asian NICs, but was pursued by the BOI in an incoherent, inefficient manner and for too long (Ingram 1971). In fact the Thai technocrats were aware of this shortcoming in the early 1960s. Nevertheless, strong pressure to retain the apparatus of ISI might have come from nationalist and populist elements within the military, manufacturers and powerful new industrial and banking conglomerates. The idea of outward-oriented trade policies was being discussed publicly in the National Economic and Social Development Board (NESDB) in the late 1960s. ISI strategy was perceived to be eventually fatal or at best self-defeating without liberalization of industrial policy and conversion to export competitiveness.

The government showed an initial indication of moving from ISI to EOI in 1972. Export incentives provided by the BOI aimed to offset the cost-increasing effect of protection on the domestic prices of intermediate goods, regarded as protection offsets insofar as exports have been able to refund the full duties and business tax on import inputs since 1972.[4] Duty drawbacks in the 1970s and the BOI's export incentives seem to have been ineffective largely because of poor administration (Akrasanee 1980). In the late 1970s, Thailand was adversely affected by the second oil shock and the subsequent world-wide recession, partly because the country had become quite open to the world economy. Balances of payments were in deficit for five consecutive years from 1975 to 1979, while the rate of inflation jumped to double-digit figures and accelerated to 19.7 per cent in 1980. However, Thailand did not rush into export liberalization immediately, not because the technocrats were wise, but because the red tape within the bureaucracy made it impossible. Thus, the gradual shift to EOI was not the result of good planning. Rather, it was obstructed by the bureaucratic state because its official regulations and procedures were superfluous and cumbersome. The instruments of import protection existed simultaneously with those of export promotion, although a given set of policy interventions that systematically promoted ISI or EOI might not have been in force. Nonetheless, the revision of the Investment Promotion Law in 1972 was designed to offset the disincentive effects of import protection.

Domestically, the middle class began to make its voice heard in politics: the business class, and ethnic Chinese business groups in particular, started to assume a more explicit role in policy-making after 14 October 1973.[5] The 1973 student uprising significantly affected the Thai political economy as a whole, and the system of patronized capitalism under the authoritarian regime was partly demolished. In other words, the parasitic relationship of clientelism seemed to have been weakened with time and the independent private sector was able to run business under the impersonal relationship of market forces. However, the patron–client bind is still very powerful and not dissolved. The military never regained the same level of unity or political dominance, despite the overthrow of the civilian government in 1976 and the establishment of a new regime in 1977 (Boonmi 1988; Chamarik 1981). Between 1979 and 1981, when the OPEC countries raised oil prices dramatically, the government could not carry out its macroeconomic policy. The economy entered a period of stagflation, experiencing for the first time twin deficits. Several austerity measures urgently adopted by the government resulted in slower growth. This economic recession partly came from the fact that the exports of primary products did not continue to earn foreign exchange. In fact, policy-makers had decided to adjust their strategy and put greater emphasis on promoting manufactured exports in the third five-year plan (1972–76). Moreover, the country shifted to an outward-looking strategy, again with the lack of a coherent industrial policy. The main stated objective to promote manufactured exports rested heavily on foreign direct investments and transnational corporations.

Nevertheless, in the fourth five-year plan (1977–81), the export promotion policy was significantly revised to reduce the anti-export bias resulting from ISI. The BOI still played a leading role in authorizing and granting exemptions and privileges. Fiscal deficits began to soar, sustained by the newly-found access to foreign commercial banks. Large-scale foreign indebtedness started in 1976; the Defence Loans Act enabled the government to borrow more for defence purposes, subject to a limit of 20,000 million baht.[6] A conservative Thai monetary policy seemed appropriate in the 1950s and the 1960s, but in the late 1970s the dollar began to appreciate against other major currencies and a fixed baht/dollar parity became untenable.

Economic policy reforms, industrialization and poverty

By 1970, the Thai government had begun to reassess its commitment to ISI in the light of growing financial difficulties. It was clear that the majority of the promoted firms' products were aimed at the urban enclave market rather than the mass rural market (Richter and Edwards 1973). Over time, the infant industry argument became increasingly evident (Akrasanee and Atjanant 1986). However, the industrialization process seems to have acquired little new technology, owing to the lack of skilled labour (UNIDO 1992). Promoted firms were required to export only a certain share of their output to receive promotion support, but the lack of effective monitoring and information to enforce performance criteria made industrial policy less effective. As a result, export-oriented manufactured goods were at best footloose. In addition, the slowdown of the world economy and reduction of American economic and military aid presented Thailand with widening balance of payments and budget deficits which made the situation worse. Even though exports had been fairly diversified since the late 1950s, the country was still characterized by primary products. The small size of the domestic market with consequent low effective demand could not absorb the excess capacity of the manufacturing sectors. It should also be pointed out that the poor and undeveloped agricultural sector has been both a cause and a consequence of this pattern.

By the early 1980s, the economy faced a set of economic crises similar to the ones that hit other developing countries. The technocrats had to undertake a major change in economic policy, especially through Structural Adjustment Loans programmes (SALs) actively advocated by the World Bank and Stand-by-Arrangements supported by the IMF (International Monetary Fund). Together with these institutions the technocrats agreed to shift the emphasis towards EOI. The objective was not only to increase exports but also to reduce balance of payments deficits and scale down the import-competing industries. SALs required major tax reform to raise more revenues and to make the tax system more efficient. In general, if structural adjustment refers to the restoration of equilibrium, i.e. providing a firm foundation to withstand further shocks and facilitate development (Goldin and Winters 1992), then SALs in Thailand during the 1980s appeared to be less satisfactory for a package of economic reform.

Ironically, Thailand cannot be classified as being successful in the implementation of at least three important policies, i.e. trade, tariffs and tax policies.[7] Furthermore, most *ad hoc* sectoral initiatives continue to prevail. Trade policies are still full of quantitative restrictions, but a clear plan for a subsequent reduction of tariffs to quite low and uniform levels was strongly advocated by the Ministry of Finance in the early 1990s (Richupan 1990). The early 1980s economic policy reforms of raising energy prices and devaluing the baht appear to have only partly achieved the desired effect. The growth of manufactured exports has increasingly been the result of sound exchange rate policy rather than export subsidies alone. Two successful devaluations in 1981 and 1986 did substantially help reduce the trade deficit in 1985 without affecting domestic levels of inflation (Ranis and Mahmood 1992). Devaluations were used to reduce the anti-export bias, and to achieve competitiveness of tradable goods. In the Thai context, the BOI has often strictly followed a one-way route by granting subsidies and privileges to both ISI and EOI. Promoted firms that did not fulfil the conditions were never threatened with sanctions.

Given the conservative financial policy of the Bank of Thailand, the policies were undoubtedly regarded as major economic reforms. Clearly, disturbances in the world economy urged Thai monetary managers to depart from the traditional exchange rate course. The volatility of the world financial system in the 1970s and 1980s also caused these adjustments, if seen in contrast to the stable period of the 1960s. Nevertheless, an exchange rate policy is not a commercial policy *per se*. This distinction is relevant in the Thai context with regard to *ad hoc* sectoral initiatives. Commercial policy affects import-substituting and export-oriented interests (Rodrik 1992). Bureaucratic autonomy, discretion and patronage have been shaped or guided by incoherent courses, in particular by the Ministry of Commerce. In 1982, Thailand became a contracting party to GATT (General Agreement on Tariffs and Trade); it was widely perceived that a commitment to integrate into the world economy would incite a country to keep pace with sound macroeconomic policy, but the contention here is that this does not necessarily turn the Thai example into a developmental state.

In essence, the lack of consistency in *ad hoc* sectoral policy (Grindle and Thomas 1991) can be attributed to the characteristics of the rent-seeking society prevailing in Thailand since the 1950s. The most important hindrance is the discretionary legal mandates of individual departments. Ministerial discretion in trade and industrial policies, quotas, licensing and factory promotions are often deployed to seek economic rents to such an extent that some of these are, or have to be, kicked back to the bureaucrats and their political masters. The trend has been increasing since the 1990s, and vote-buying politics might contribute to it as well. It should be made clear that the East Asian NICs may have experienced rent-seeking to a considerable extent, but the social benefits of high productivity in agriculture and export-led industrialization exceeded the social costs of rent-seeking activities.

It is evident that export licensing, either automatic or non-automatic, affects about 150 product categories, mainly textiles and clothing, which have rapidly become the country's highest foreign exchange earning commodity group. They also contributed substantially to value-added manufacturing and absorbed approximately three-quarters of employment, especially female workers in that sector. Other items cover certain agricultural commodities, fuels, metal and metal products, wood and wood products, wild animals and their carcasses, pesticides, paper and sacred statues and images. Export quotas are still in place for sugar, cassava and textiles. Formal institutions, in particular the Import and Export Commodity Act of 1979, give absolute power to the Ministry of Commerce or the Permanent Under-Secretary of the Ministry to promulgate scores of subordinate laws for imposing quantitative restrictions and other regulations on trading without the approval of the Cabinet or Parliament (Siriprachai 1990). This can easily lead to corruption and rent-seeking activities if the bureaucrat is not a benevolent social guardian. In the Thai context, the bureaucrat can restrict supply with very low risk of detection or punishment from above. Rather, clientelism encourages corrupt officials to expand their activities. It is very common for their bosses to share in the on-going process (Shleifer and Vishny 1993). The low salary is often cited as the reason for this, but the situation might be more complicated in the sense that the institutions cannot keep the most honest and talented people within their organizations. What is more important, Thailand has never had administrative courts to deter corrupt bureaucrats. Such an institution would serve to curb malfeasance, impropriety and abuse of power on the part of state officers (Klausner 1989). Hence, legal and public cessation of corruption and rent-seeking is highly unlikely to come about.

In addition, under the existing electoral regime and elected coalition government after 1975, politicians who were to be re-elected and desired to be ministers needed to spend a lot of money on patronage to keep themselves in office. Clientelism became pervasive. It is often stated that trade quotas, capacity control and factory permits created resources for the military elite in the past, but this applies to political parties and individually elected politicians in the present. Furthermore, the serious shortage of infrastructure since the middle of the 1980s has created an opportunity for the various ministries (elected politicians) to carry out big projects. Economic rents (in the form of commissions, permits, licences, etc.) were repeatedly appropriated by bribery, 'palm-oiling' or other corrupt means. However, in many cases the allocation system is not based on competitive rent-seeking, but is used covertly instead by the politicians in power to generate income for themselves or for their party. Hence, commercial policy, to a greater extent than that of *ad hoc* sectoral policies in Thailand, has seldom been reformed through the SALs.[8]

It has been argued that both SALs and the Stand-by Arrangement programme in the early 1980s rescued the Thai economy, which readjusted towards greater efficiency-enhancing measures at the macroeconomic level; but this may have been too little and too late to encourage balanced growth and equal distribution of income and opportunities for the rural poor and unskilled labour. Several studies have shown that the nominal exchange rate of the baht had been overvalued for years (Siamwalla and Setboonsarng 1987), due to the imposition of high import duties to protect promoted firms. In other words, the agricultural sector was squeezed by the modern sector for a long period and as a consequence, Thai peasants had to pay the high costs of the protected goods. It is manifest that both *ad hoc* sectoral policies, for example, in rice, and an overvalued exchange rate during the 1961–80 period, conspicuously discriminated against export producers; in particular exporters of primary commodities. Not surprisingly, agriculture, which is the poorest sector in the country, has been left in poverty. As rightly observed by Timmer:

> Thailand did not use similar trade and pricing for key commodities in an effort to protect domestic farmers from the very low prices that occur from time to time in the world market. Although the strong performance of Thailand with rising labour productivity argues that such free trade policies promote growth, Thailand paid a price for rural poverty. (Timmer 1991: 138)

During the readjustment in the 1980s, the Thai government seemed reluctant to implement the SALs fully. Development policy aimed instead at alleviating poverty in the agricultural sector, and government expenditures as well as the tax system were moved firmly towards the creation of an environment suitable for EOI industries, especially around Bangkok. It is often forgotten that in the NICs, the domestic market was a crucial prerequisite before manufactured export goods could compete on the world market (Gunnarsson 1991). This fact might not have been discussed among the technocrats in the NESDB.

It is also obvious that the scale of land reform was very limited and ineffective in preventing poor farmers from becoming indebted. Until the Agricultural Reform Act of 1975, attempts to limit private ownership were never successful due to vested interests. The revision of land reform in 1993 under the Chuan government is a good example in support of this assessment. The agrarian reform scandal in many provinces, particularly in Phuket in late 1994, suggests that discretionary power and rent-seeking activity were deployed as instruments to seek economic advantages from scarce resources, while the state still strongly supported the policy of growth maximization. The equalization of regional and personal income levels is believed to have received even lower priority than had been expected during the 1980s.

Of course the impressive growth since the 1960s has to some extent trickled down some benefits to the poor. Absolute poverty declined steadily from 57 per cent in 1962–63 to 24 per cent in 1981. But worsening income distribution in Thailand over three decades has remained ubiquitous (Huntaserini and Jitsuchon 1988; Tinakorn 1992). The income shares of the richest 20 per cent of households increased from 50 per cent of total household income in 1975/76 to 55 per cent in 1988/89, while the share of the poorest 20 per cent declined from 8 per cent to 4.5 per cent during the same period (Sussangkarn 1992). However, recent data show that the share of the richest 20 per cent of households decreased to 42.81 per cent in 1990, while the share of the poorest 20 per cent jumped to 8.52 per cent.

Rising income inequality both between industrial and agricultural sectors and between regions partly reflects the nature and competence of the state to cope with fundamental problems. It remains apparent that rising income equity and real wages in the rural areas cannot be implemented through a laissez-faire policy. In addition, both import-substituting and export-oriented industrial sectors, mostly concentrated in Bangkok (Santikarn 1992), have been inadequate for sufficient labour absorption. An attempt in the 1980s to restructure the economy appears to have offered little stimulus to industrial labour absorption. In addition, the output elasticity of employment in the manufacturing sector was very low and declining in the 1980s compared to the 1970s; falling noticeably from 0.74 to 0.39 (Ghose 1993, Table 4.11). Employment conditions almost certainly deteriorated regardless of the development strategy adopted. It has been argued that export-oriented industries are significantly more labour-intensive than import-substituting industries, but this was not the case for Thailand.

What factors can explain the very low elasticity of employment in the manufacturing sector in the 1980s? The monsoon weather also affects seasonal fluctuations in labour demand. In the long slack season, men and women migrate to work in Bangkok, partly because there are not enough jobs in rural industries. Having only low skills and education (due to a weak and predatory state), they often worked in informal sectors and lived in the slums of Bangkok. The formal sector, led by transnational and joint-venture firms, did not have sufficient capacity to absorb labour. Migration provided essential means of escape from poverty as the migrant workers could only get jobs in the informal sector with its easy entry and low skills requirements. Small income earnings from hard work were deployed to support families in their rural homes. The push factor was of greater importance than a pull-type urbanization since it was the deficiency of employment opportunities stemming from the collapse of the rural economy which forced agricultural workers to migrate to Bangkok.

By the middle of the 1980s, commodity prices on the world market were depressed and a substantial number of farmers remaining in agriculture received very low incomes. It is not surprising that the price support programmes hardly ever had adequate funds to raise the general level of farm-gate prices for rice (Thanapornpun 1980). In the case of cassava, the quota system causes a large source of rent-seeking and rent-dissipation (Siriprachai 1988); the exception is sugar-cane, where the mills' and farmers' associations are strong and can cooperate in bargaining with the government. The successful structure of subsidy in the sugarcane industry is due to the nature of the production line.

It is commonly known that the Thai economy has undergone rapid economic transformation as the share of agriculture in the GDP has been shrinking rapidly and as export promotion has replaced import substitution. However, the tax system has remained very regressive compared with the NICs (Tanzi and Shome 1992) and has lacked transparency. The tax structure itself is rather complex, resulting from many special allowances and high standard deductions (allowed for different sources of income) granted in particular by the BOI, and from the failure to tax fringe benefits. Tax policy has historically depended heavily on domestic consumption and trade levies. By 1992, the 7 per cent value-added rate was confusedly replaced by complex, inefficient business taxes by the Anand government. Thailand has tinkered with its tax system over the years without any major policy reform (Patamasiriwat 1993). The existing policy does not provide for an equitable income distribution. It is strange that while the state has played a key role in enhancing economic growth, tax policy has not served to generate social justice and human rights (Tinakorn 1992). Although the signal that the government has put out over the years to stimulate exports of manufactures is still effective, income redistribution or the achievement of special social goals is a long way off. Not surprisingly, the role of social security with transfer payments has been very limited (Tanzi and Shome 1992). It is evident that levies to discourage speculation in land have never been put forward. An unintended consequence for the urban poor is that their rents and therefore general living costs increase accordingly.

Commercialization has gradually taken place since the 1980s and increased the numbers of vulnerable agricultural wage-earners and unskilled workers on the verge of becoming poor (Tinakorn 1992), partly because the land frontier has been closed. In addition, recent expeditious high land prices have given rise to unproductive expenditures and social waste arising from speculation, while tax incentives granted by the BOI have led to prevalent rent-seeking by a public bureaucracy made up of members of a low-paid, poorly trained, dishonourable but powerful elite (Christensen and Siamwalla 1993). Recently, the BOI began to grant subsidies to luxury hotels and urban housing developers, supposedly to provide the urban poor with accommodation. The main objective of establishing the BOI to protect and help infant industry is fading. The social cost of distortions has always been treated as negligible, but is in fact substantial.

In contrast to Thailand it must be emphasized that the case of the NICs was quite unique. The beneficial effects of incentives will not develop if incompetence, corruption or various forms of rent-seeking activities become the norm. Since the late 1950s, incentive legislation, especially in *ad hoc* sectoral policy that is based on discretionary decisions, has provided a perfect instrument for enriching a few members of the military elite, bureaucrats and elected ministers by permitting some investors or traders to evade taxes or to get quotas or licences. The loser is, of course, the public interest.

The tariff reform in the 1980s seems to have had little effect. The Thai government will need to focus on reducing the wide dispersion of its 34 nominal tariff rates ranging between one and 200 per cent and the excessive rates of effective protectionism. In fact, a major tariff reform was attempted in 1982 but although some progress was made in equalizing rates of protection within individual industries such as textiles, the result for the industrial sector as a whole was modest. In effect, some of these gains were reversed when tariff rates were raised again in 1985. The import duty was still essential for increasing government revenue when the budget remained in deficit. This left Thailand with the highest average nominal rate of tariff protection (34 per cent) of any of the Pacific Basin developing countries (Noland 1990). The government again began to reform tariffs in the early 1990s as a result of a commitment to the ASEAN Free Trade Area (AFTA) to do so within 15 years. Export duties have been of less importance since the 1980s, while import levies have remained high at 18 per cent of government revenues.

The state has also been incapable of dealing with privatization. National or security interests are often cited as reasons to keep the loss-making state-owned enterprises. Both civilian and military governments alike have distributed economic patronage by appointing their supporters and friends to control state-owned enterprises. In other words, the strong clientelistic relationships have been strengthened and used to seek economic rents to finance political parties in the next general election.

The early 1980s witnessed a slowdown in infrastructure investment in a Thailand facing twin deficits. Most government projects were postponed to keep the economy stable. Nevertheless, after the mid-1980s the economy rapidly recovered. The state has been inviting foreign private enterprises to invest in huge infrastructure projects by granting concessions on a build-operate-transfer basis. It stems in fact from the ideology of privatization. It is interesting to mention that it has undoubtedly led to a rent-seeking war among elected politicians, government officials, state enterprises and private investors, because each mega-project can easily constitute a gigantic income for those who can exert the greatest political muscle (Mueller 1989). A simple explanation is that privatization started without any attempt to draft a standardized legal framework to regulate it. Hence, bribery and 'palm-oiling' have been flourishing. In the Thai context, it is commonly known as 'tea' or 'coffee-money'.

The Thai state: between rent-seeking and inequality

Thailand has a number of administrative laws that concern the implementation of various aspects of national policy, especially in international trade and internal affairs. Political pressures and vested-interest groups such as exporters, industrialists and bankers have exerted influence, while workers and peasants are less powerful and unable to bargain with bureaucrats and vested-interest groups. The fragmented political party systems and vote-buying remain evident. Political parties receive financial support from urban vested-interest groups and are without any significant bases in the countryside. In fact, bureaucrats and ministers usually find it prudent to consult formal interest groups – namely trade, industrialist and banker associations – on various technical aspects of state intervention. As pointed out by Laothamatas (1992), this has provided these vested-interest groups with the means of entry into the policy-making process, whereas other societal groups are never informed of the consequences of state discretionary policies. However, some trade associations are more skilful in lobbying policy-makers than others. Not surprisingly, inside information has allowed interest groups, for instance exporters, to obtain large speculative profits through their trading activities. It is worth noting that, since 1973, unstable governments have tended to concentrate resources in urban and industrial sectors with the aim of maintaining their own power and winning international recognition. Previously, rural insurgency in the northeast in the 1960s and the 1970s had compelled the elite in Bangkok to pay more attention to the rural poor and distribute some resources.

Reliable and accurate data on the standard of living is notoriously difficult to collect, but trends in poverty and labour productivity within the rural sector reflect the underlying economic environment in which these people live and work. The rural poor, not just in the northeast but also in other regions, are very destitute and have had only little education, usually discontinuing their studies after the compulsory level. It is fair to say that part of the infrastructure development by the central government in remote rural areas helped to improve the living conditions. Nevertheless, the economic surplus in the rural sector has been squeezed and used to subsidize urban industrial development; the rice premium was one such case between 1955 and 1986. After enacting the Farmers' Aid Fund, the government still kept the domestic prices of rice down through rice reserve requirements adopted by the Ministry of Commerce (Thanapornpun 1985). This inexorably led to a drop in the cost of living in urban areas, reduced the pressure by urban workers for wages increases, and enabled export-oriented promoted firms to compete in the world market.

If industrialization in urban areas had been able to bring about modernization in agriculture, it would have been possible for Thailand to become an NIC. But this failure is, to a large degree, due to the industrial strategy. Again, the East Asian countries have grown at a very high rate and promoted high-technology industries, presumably by tax incentives under a strong activist and developmental state. The essential difference is that the Thai state has not come out with threats to punish promoted industries for not being competitive or efficient. Furthermore, the technology transfer process, being one of the most important linkages of late industrialization, is the missing link of industrial policy. The nationalist policy during the Phibul era might be close to the insulated developmental state in South Korea (Haggard and Moon 1983); but under the bureaucratic capitalism of state-owned enterprises, it turned to an import-substitution strategy in which inefficient industries needed to be protected from foreign competition for many years. The redirection of policies reflects an emerging consensus that countries following an outward-oriented, market-based development strategy achieve relatively higher rates of growth and living standards. It may be too early to confirm this in the case of Thailand.

In the present conditions, there are still some difficulties and restricting conditions that must be remedied to secure continued growth of the Thai economy. One immediate problem is bottlenecks in the infrastructure and another is the quality of the labour force. Infrastructure investment needs are enormous, but in part because of large government budget deficits in the early 1980s that surpassed 6 per cent of GDP, Thailand has saved and invested somewhat less than some other East Asian countries. Even though the government has run a surplus since 1988, the share of investment in GDP has risen only to about 26 per cent. Additional increases to around 30 per cent would be desirable in the long run (Noland 1990). However, since the early 1990s this has gradually increased to approximately 30 per cent which was made possible by high sustained economic growth.

As a result of corruption in government contracting, the flow of investment in infrastructure has been adversely affected. Additional investment in human capital is also urgent, including an expansion of secondary education in general, and of technical and vocational education in particular. Sciences and engineering need to be strengthened at university level, research and development activities encouraged domestically. The main point is to create equal opportunities for the lower classes. The consecutive surpluses in government budgets make it a propitious time for the government to carry out major reforms.

For the period from the 1960s to the 1980s, data tells us that the trade-off paradigm between the speed of economic growth and the share of income of the poor was so little mitigated that the deterioration in the share of income of the poor was quite substantial. The conclusion is that the path of development does not give much hope in the foreseeable future of attaining the goal of poverty eradication through economic growth (Adelman 1992).

Thailand may have been able to achieve stabilization in macroeconomic policy, and in particular high economic growth, a low inflation rate and a surplus in government budgets. According to the World Bank this is due to the outward-looking strategy, for which export expansion has been the engine of recent sustained economic growth (1993). Structural adjustment aimed at correcting imbalances in foreign payments, government budgets and the money supply with the aim of controlling inflation and maintaining macroeconomic stability is obviously successful. On the other hand progress in economic reform, liberalization, privatization and institutional overhaul is questionable. Major institutional and administrative reforms were hardly ever carried out. Recent attempts at Constitutional Amendment in late 1994 on the issue of devolution of authority to local government have failed entirely. Liberalization through the removal of government interventions in *ad hoc* sectoral policy, namely incentives for investment, export quotas, export licensing and other barriers to entry were not seriously implemented either. Such a policy is known to foster patronage and rent-seeking, rather than to dismantle it. A case in point is the Commerce Ministry corruption scandal in allocating quotas of cassava to the European Union in 1993–94. Privatization through the sale of government-owned enterprises and the contracting of formerly governmental functions was negligible in the past.[9] Institutional reforms, in the sense of changes in government institutions that make it possible for economic reform to work and predominantly involve shifts away from administered control towards mechanisms that reduce transaction costs of administration, have been rare.

In conclusion, the economy needs reform. Rent-seeking and corruption should be kept at a minimum by making administrative regulations clear, transparent and accountable. In fact, Thai society absolutely needs good governance and in particular sound development management (Leftwich 1995). Accountability, a legal framework for development, public information and transparency should be fully provided to warrant democracy. The lack of a mechanism for effectively controlling and countering discretionary power is a serious drawback in a regime of patron–client relationships which might be considered the major barriers to reform. As in other countries with well-developed clientelistic governments, economic reform can be both politically costly and irrational.

Most neo-classical economists seem to conceive that export-oriented policies reduce inequality within developing countries, but the Thai case appears not to have confirmed this proposition. Thus, more expansion of manufactured exports into the world market does not necessarily lead to reduction of inequality. The last point that should be noted is the institutional constraint that shapes Thai society into being neither a strong nor minimal interventionist state. The predatory-cum-soft authoritarian state might be an apt description of Thailand.

1. I am very grateful to Laurids S. Lauridsen, Christer Gunnarsson and Johannes D. Schmidt for helpful comments and suggestions. Jaya Reddy has done an excellent job in making my language accurate and readable.

2. The prestige is given to academics or experts (Nak Wichakarn). Thai technocrats often include professionals/experts working at the NESDB and the Bank of Thailand or the Ministry of Finance.

3. Siamwalla and Setboonsarng (1987) rightly observed that the BOI emphasized the promotion of big industry with tariff protection, exemption from import tariffs on machinery and income tax holidays. Unfortunately, most agro-processing industries like rice milling and rubber processing (small and medium-scale) are entirely ignored in terms of access to BOI promotion privileges.

4. In fact, this system was established in the late 1950s under the control of the Fiscal Policy Office of the Ministry of Finance, but the law allows for a partial refund of duties and business taxes on inputs.

5. This was a popular uprising of university students against the military government which resulted in the downfall of the Thanom-Prapath regime (the Sarit's legacy).

6. In 1981 the Act was revised to empower the government to borrow from foreign sources for defence purposes as long as the sum of defence borrowing plus external borrowing for other purposes did not exceed 10 per cent of the budget expenditure each year.

7. The World Bank classifies the conditional content of structural adjustment loans programmes in ten categories: exchange rate, trade policies, fiscal policies, budget/public spending, public enterprises, financial sector, industrial policy, energy policy, agricultural policy and other (see Greenaway and Morrissey 1993).

8. The main policy recommendations of the World Bank for Thailand were: (i) to raise domestic energy prices to international level; (ii) to develop strong deflationary monetary and fiscal policies; (iii) to end the import substitution policy for industry; (iv) to emphasize export-oriented industries; (v) to reduce import tariffs and remove all export restriction and taxes; (vi) to increase personal taxation and make it more effective; (vii) to end restrictions on the level of domestic interest rates; (viii) to undertake a comprehensive review of government organization and expenditure in order to eliminate waste, etc.

9. There were indeed two state enterprises sold to private firms during the 1977–91 period, four expired, two were rented by private enterprises, one received a concession, four became joint-ventures, and four were permitted into the private sectors to run businesses. There were 61 state-owned enterprises at the end of January, 1993.

Bibliography

Adelman, Irma (1992) 'What is the Evidence on Income Inequality and Development?' in D. J. Savoic and Irving Brecher (eds) *Equity and Efficiency in Economic Development: Essays in Honour of Benjamin Higgins*, Mcgrill Queen's University Press, Quebec

Aeusrivongse, Nidhi (1993) 'On the Future Road', A Synthesis Report Volume I, paper presented at the 1993 Thailand Development Research Institute year-end conference: Who Gets What and How?: Challenges for the Future, 10–11 December, Pattaya, Thailand (in Thai)

Akrasanee, Narongchai (1980) *Industrial Development in Thailand. Report of the Research and Planning Department*, IFCT, Bangkok

Akrasanee, Narongchai and **Juanjai Atjanant** (1986) 'Thailand: Manufacturing Industry Production Issues and Empirical Studies', in C. Findlay and R. Garnaut (eds) *The Political Economy of Manufacturing Protection: Experiences of Asean and Australia*, Allen & Unwin, London

Amsden, Alice H. (1989) *Asia's Next Giant: South Korea and Late Industrialization*, Oxford University Press, New York

Boonmi, Thirayuth (1988) 'Thailand: A Political Figuration of the Traditionalistic Controlled Bourgeois Representation', *Social Science and Humanities Journal*, **15**(2), (September–December), 46–67

Chaloemtiarnana, Thak (1979) *Thailand: The Politics of Despotic Paternalism*, Thammasat University Press, Bangkok

Chamarik, Saneh (1981) 'Problems of Development in Thai Political Setting', paper presented at the First International Conference on Thai Studies, 25–27 February, 1981, New Delhi

Christensen, Scott and **Ammar Siamwalla** (1993) 'Beyond Patronage: Tasks for the Thai State', paper presented at the 1993 Thailand Development Research Institute year-end conference: Who Gets What and How?: Challenges for the Future, December, Pattaya, Thailand

Christensen, Scott and **Ammar Siamwalla** (1994) 'Muddling Toward an Economic Miracle', *Bangkok Post*, Thursday, 23 June, p. 18

Evans, Peter (1995) *Embedded Autonomy: States and Industrial Transformation*, Princeton University Press, Princeton, NJ

Ghose, Ajit K. (1993) 'Global Changes, Agriculture and Economic Growth in the 1980s: A Study of Four Asian Countries', in Ajit Singh and Mamid Tabatabai (eds) *Economic Crisis and Third World Agriculture*, Cambridge University Press, Cambridge

Goldin, Ian and **L. Alan Winters** (1992) *Open Economies: Structural Adjustment and Agriculture*, Cambridge University Press, Cambridge

Greenaway, David and **Oliver Morrissey** (1993) 'Structural Adjustment and Liberalisation in Developing Countries: What Lessons Have We Learned?', *Kyklos*, **46**, 241–61

Grindle, Merilee and **John W. Thomas** (1991) *Public Choices and Policy Change: The Political Economy of Reform in Developing Countries*, Johns Hopkins University Press, Baltimore

Gunnarsson, Christer (1991) 'Dirigisme or Undistorted Free-Trade Regimes?: An Historical and Institutional Interpretation of the Taiwanese Success', paper presented at the Arne Ryde Conference on International Trade and Economic Development, June, Elsinore, Denmark

Haggard, Stephen and **Moon Chung-In** (1983) 'The South Korean State on the International Economy', in John G. Ruggie (ed.) *The Antinomies of Interdependence*, Columbia University Press, New York

Huntaserini, Suganya and **Somchai Jitsuchon** (1988) *Thailand's Income Distribution and Poverty Profile and Their Current Situation*, paper presented at the Thailand Development Research Institute year-end conference, Pattaya, Thailand

Ingram, James C. (1971) *Economic Change in Thailand, 1970–1980*, Stanford University Press, Stanford, California

Klausner, William J. (1989) 'Thai Society's Legal Barriers', *Solidarity*, **121**, (January), 57–64

Laothamatas, Anek (1992) *Business Associations and the New Political Economy of Thailand: From Bureaucratic Polity to Liberal Corporatism*, Westview Press, Boulder, Colorado

Leftwich, Adrian (1995) 'Bringing Politics Back In: Towards a Model of the Developmental State', *Journal of Development Studies*, **31**(3), (February): 400–27

Marzouk, G. A. (1972) *Economic Development and Policies: Case Study of Thailand*, Rotterdam University Press, Rotterdam

Meesook, Oey Astra, Pranee Tinakorn and **Chayan Vaddhanaphuti** (1988) *The Political Economy of Thailand's Development: Poverty, Equity and Growth, 1850–1985*, research report submitted to the World Bank

Mueller, Dennis (1989) *Public Choice II*, Cambridge University Press, Cambridge

Noland, Marcus (1990) *Pacific Basin Developing Countries : Prospects for the Future*, Institute for International Economics, Washington, D.C.

Patamasiriwat, Direk (1993) 'Tax Reform', paper presented at the XVI Annual Symposium on Fiscal Revolution for Economic Policy Reform in Thailand, held by Faculty of Economics, Thammasat University, Bangkok, Thailand (in Thai)

Ranis, Gustav and **S. A. Mahmood** (1992) *The Political Economy of Development Policy Change*, Basil Blackwell, Oxford

Richter, H. V. and **C. T. Edwards** (1973) 'Recent Economic Development in Thailand', in R. Ho and E. C. Chapman (eds) *Studies of Contemporary Thailand*, Australian University Press, Canberra

Richupan, Somchai (1990) 'Tax Policy and Economic Development in Thailand', paper presented at the Conference on Tax Policy and Economic Development among Pacific Asian Countries, Taipei (January)

Riggs, Fred W. (1966) *Thailand: The Modernisation of a Bureaucratic Polity*, East–West Centre Press, Honolulu

Riggs, Fred W. (1993) 'Bureaucratic Power in Southeast Asia', *Asian Journal of Political Science*, **1** (June): 3–28

Rodrick, Dani (1992) *The Rush to Free Trade in the Developing World: Why*

So Late? Why Now? Will It Last?, Working Paper No. 3947 (January), National Bureau of Economic Research

Rosenberg, Hans (1958) *Bureaucracy, Aristocracy and Autocracy: The Prussian Experience, 1660–1815*, Harvard University Press, Cambridge, Mass.

Samudavanija, Chai-anan (1992) 'Industrialisation and Democracy in Thailand', paper presented at the conference on the Making of A Fifth Tiger? Thailand's Industrialization and Its Consequences, December, Australian National University

Santikarn, Kaosa-ard (1992) 'Manufacturing Growth: A Blessing for All?', Thailand Development Research Institute Year–End Conference, Pattaya, Thailand

Shleifer, Andrei and **Robert W. Vishny** (1993) 'Corruption', *Quarterly Journal of Economics*, **CVIII**(3) (August), 599–617

Siamwalla, Ammar (1993) 'Four Episodes of Economic Reform in Thailand, 1958–1992', Thailand Development Research Institute, mimeograph (July)

Siamwalla, Ammar and **Suthad Setboonsarng** (1987) *Agricultural Pricing Policies in Thailand, 1960–1985*, report submitted to the World Bank (October)

Siriprachai, Somboon (1988) *VER and Thai Government Policy Implementation: the Case of Cassava Trade between the European Community and Thailand*, M.A. Thesis, Faculty of Economics, Thammasat University (in Thai)

Siriprachai, Somboon (1990) *Thai Law and International Trade Sectors: A Case Study of Rice and Cassava Exports*, Research report submitted to Faculty of Economics, Thammasat University (February) (in Thai)

Siriprachai, Somboon (1993) *Can Southeast Asian States Emulate East Asian Developmental States?*, NONESA (Newsletter of the Nordic Association for Southeast Asian Studies), No. **8**, 9–16

Siriprachai, Somboon (1995) 'Population Growth, Fertility Decline, Poverty and Deforestation in Thailand, 1850–1990', in Mason Hoadley and Christer Gunnarsson (eds) *Village in the Transformation in Rural Southeast Asia*, Curzon Press, London

Sussangkarn, Chalongphob (1992) *Towards Balanced Development: Sectoral, Spatial and Other Dimensions*, Thailand Development Research Institute Year-End Conference, Pattaya, Thailand

Tanzi, Vito and **Parthasarathi Shome** (1992) 'The Role of Taxation in the Development of East Asian Economies', in Takatoshi Ito and Anne O. Krueger (eds) *The Political Economy of Tax Reform*, University of Chicago Press, Chicago

Thanapornpun, Rungsan (1980) *The Role of Farmers' Aid Fund*, Research

Report Series in Agricultural Pricing and Marketing, submitted to NESDB (in Thai)

Thanapornpun, Rungsan (1985) *The Economics of Rice Premium: Limits of Knowledge*, report submitted to Thai Khadi Research Institute, Thammasat University (in Thai)

Thanapornpun, Rungsan (1990) *The Process of Economic Policy Making in Thailand: Historical Analysis of Political Economy, 1932–1987*, Social Science Association, Bangkok (in Thai)

Timmer, C. Peter (ed.) (1991) *Agriculture and the State: Growth, Employment and Poverty in Developing Countries*, Cornell University Press, Ithaca, NY

Tinakorn, Pranee (1992) 'Industrialization and Welfare: How Poverty and Income Distribution Are Affected', paper presented at the conference on the Making of A Fifth Tiger? Thailand's Industrialization and Its Consequences, December, Australian National University

UNIDO (1992) *Thailand: Coping with the Strains of Success*, Basil Blackwell, Oxford

Wade, Robert (1990) *Governing the Market: Economic Theory and the Role of Government in East Asian Industrialization*, Princeton University Press, Princeton, NJ

World Bank (1993) *The East Asian Miracle: Economic Growth and Public Policy*, Oxford University Press for the World Bank, Oxford

Societal forces and class constellations behind Southeast Asian capitalism

Johannes Dragsbaek Schmidt, Jacques Hersh and Niels Fold

During the two decades preceding the 1990s, the notion of classes was a pivotal aspect of political and sociological analysis before almost fading away in recent years. It seems as if this marginalization of the class concept in social research was the outcome of, and response to, both the crisis of Marxism in the 1980s and the dissolution of 'real existing socialism'. This is not surprising when taking into consideration the subjective factor of social sciences as well as the problem of research persistently lagging behind events happening in the 'real world'. In the 1970s, the theory and method of dialectical materialism and the Marxist critique of capitalism had an almost hegemonic discursive status in the 'sociology of Zeitgeist', not only in development studies but also in social research in general. Marxism appeared already to have come to an impasse in the mid-1980s due to a generalized theoretical disorientation and the lack of openness to diversity (Booth 1985; 1994). Moreover, research within critical social change perspectives and Marxist studies became confronted with a trend towards renewed empiricism and with intellectual positions stemming from postmodernism and postcolonialism (Corbridge 1994: 90; 112 fn. 2).[1]

But as the contributions to this volume show, albeit with diverse and critical theoretical points of departure, social forces did not die together with the fading away of class as a social science construct and the demise of Soviet-style Marxism-Leninism. On the contrary, globalization and uneven development have put renewed focus on social actors such as classes and state entities, thus underlining the importance of a non-reductionist understanding of changing class formations and constellations. In order to provide adequate understandings and explanations of rapid social change, an approach incorporating a broad sociological perspective might be recommended. It follows that a shift in attention to a 'new comparative political economy' should be brought to the fore.[2] Either way, the study of classes in capitalist societies is always, by definition, based on an analysis of social inequality. Without inequality there would be no differentiation. In its most simplistic terms, a capitalist social order consists precisely of groups of people occupying common positions characterized by inequality.

Based on these observations, this concluding chapter aims firstly to provide a reappraisal of class formation in Southeast Asia, secondly to deconstruct the anomalies of the current discussions on the East Asian model and reassess its relevance for Southeast Asia and thirdly to establish whether new social forces are emerging in the region and how to identity the resulting new patterns of class formation, social change and inequality.

Class formation and the politics of late growth in Southeast Asia

Class formation in Southeast Asia has not followed any predetermined pattern. Orthodox Marxism and classical modernization theory have historically conceived the bourgeoisie as a transformative dynamo, as it concentrates and centralizes capital in ways that lead to the sharpening of contradictions between capital and labour. In Southeast Asia these characteristics have so far been overshadowed by state action although some studies have pointed to the emergence of domestic business groups (Doner 1991 and 1992; MacIntyre and Jayasuriya 1992; Laothamatas 1992).

According to these authors, Thailand and Indonesia illustrate the movement of bureaucratically dominated policies towards a recent trend of greater independence for business,[3] while Malaysia has moved in the opposite direction by imposing political controls on property and ethnic Chinese capital and promoting the nascent nationalistic state-induced *Bumiputra* bourgeoisie.[4] However, the situation is fraught with uncertainty, and it is not clear whether the promotion of large-scale Malay businessmen will reduce state strength.[5] Even in Indonesia recent moves by the state apparatus to impose restrictions on ethnic-Chinese-owned properties and curtail links between ethnic Chinese and foreign capital prove that the process of social reordering remains uncertain. The Indonesian state appears to be in a process of revitalizing nationalism and protectionism. The case of Singapore is one of extreme reliance on foreign investment, to such an extent that it has effectively marginalized the domestic bourgeoisie, both economically and politically (Deyo 1987: 234); the local capitalists in that city-state are certainly not industrialists, their strategies remain predominantly short-term and they are most often in a secondary position behind international capital (Paix 1993: 195).

In the case of the Philippines the capitalist class has several large, old-established family-controlled corporate conglomerates of a kind not readily found elsewhere in the region. Although the position of big business is much more embedded, the Philippines is also the only country in the region to have hardly seen a year pass since the late 1960s without the government having to enter either a standby-agreement or an extended fund facility agreement with the International Monetary Fund, thus placing constraints on the state's monetary and fiscal policy-making. The latest developments in the country show that the Ramos administration is attacking the major oligarchic family firms and challenging their positions in the business cartels and monopolies.

The Vietnamese situation is more complicated as it involves a double-transitionary movement of the economy resulting in what Andreff has called a paradoxical situation: the transition out of socialism and under-development, which seems to be rather efficiently handled by a Communist Party still in power but convinced of the benefits of a market economy. This takes place at a time when, in Central and Eastern Europe, economic transition is regarded as the twin sister of political democracy and flourishing parliamentary debates. The Chinese economic reforms since 1978 is probably another case in point (1993: 529). The gradual reforms of *Doi Moi* in Vietnam are reorganizing the composition of class structures and relationships, as well as the concentrations of economic power within the existing social groups. There has been a large increase in the size of the private sector, both in and outside agriculture, but the resulting social differentiation has created serious problems with regard to new divisions in both sectors (Beresford 1993: 220). The Vietnamese Communist Party has adopted the Asian NIC model of state-led industrialization and 'the mobilizing basis of the present reform process is precisely to replace class consciousness with a new economic nationalism.'(Greenfield 1994: 204) In addition:

> The increased involvement of the military in capitalist enterprise and the rise of merchant capital from within the party-state bureaucracy has created the conditions for conflict between this movement and a nascent bourgeoisie whose power remains symbiotic with the structures of state power. (Greenfield 1994: 223)

These historically diverse contingent developmental paths are good illustrations of the fact that analysis has to move beyond a point of departure of simplified Marxist exclusivistic social relations of production. It has become clear that the Weberian approach has a number of insights which cannot be left out of the analysis. In other words:

> As 'Weberians' have become worried about the Boundary problem, and 'Marxists' have recognized the importance of the middle classes, the theoretical waters were bound to become muddy. In these circumstances it is more profitable to worry less about the way in which one discourse is privileged over another and more about the manner in which adequate theory of the middle class can be constructed. (Abercrombie and Urry quoted from Thrift and Williams 1987: 1)

Thus the concept of class is not to be understood in the traditional sociological meaning as merely a structuration layer but should be comprehended in a dynamic sense, i.e. as a social relationship. A non-reductionist point of departure retains the idea of a mode or a societal relationship of domination as the organizing principle of class in differential positions. In this vein, classes gradually crystallize from latent interest groups into action groups as their interests become manifest through ideology, consciousness, leadership, and organization, and these groups become the main vehicles for conflict and change (Dahrendorf 1959 quoted from Smelser 1994: 7).

This all-encompassing approach is variously conceptualized by Robison, Kahn, Drakakis and Mulder in their contributions to this volume showing that this in-between social state has not yet emerged as a class 'in itself', but remains a highly fragmented social construct. Futhermore, history shows that when middle-class interests are threatened by powerful radical working-class movements, or when the middle class fears social chaos as in Chile, Argentina, Italy, Spain or Korea, it is prepared to support authoritarian or even fascist regimes (Rodan 1995a: 6). This is also consistent with the findings of Hewison, which demonstrate that, 'significant elements of the capitalists and "middle" classes actively support authoritarian regimes because their social position, their access to resources and their protection from other social forces, require the coercive power of the state' (Hewison et al. 1993: 6). In fact, civil society has historically coexisted with authoritarian regimes in Southeast Asia (Hewison and Rodan 1994). As also noted by Hersh in the opening chapter of this volume, what is most distinct about the middle classes in the region is the fact that they are creatures of state policies and their destiny is closely tied to the fate of the state. This is very nicely capsulated by Subangun with regard to the Indonesian case:

> The existing middle class was not born on its own but was created by state policy, and tightly connected with state activities. Thus, it is a typical process in Indonesia that under the banner of deregulation, a structural change is absent, because this type of program is strictly limited to a very formal level. It is not society which owns capital, side by side with the state-owned capital. The Indonesian middle class will fall or succeed only together with the government. (1989: 73)

Other determinants also impact on societal divisions and organization. On the face of it, these cannot be reduced to political and economic categories, but they are nevertheless part of these categories. Autonomous or relatively autonomous social forces might act within the limits described by class structure such as race, religion, ethnicity, gender, family and various state apparatuses, not only blurring the basic class divides but also generating their own social divisions (Wright 1985). Indeed, as clearly emphasized by Kahn in Chapter four, one of the most heated controversies within social sciences concerns the prerequisites for social change, not least in the Southeast Asian context. This is why he suggests that the present state of Malay neomodernity and the sophistication of the middle class cannot be analysed adequately by class analysis, but rather by a political–cultural approach. This critique of class-based analysis is further extended by Mulder, who identifies the culture of the urban middle class in the Philippines through a focus on changing identities and mentalities. He shows that the new trait of this strata is the absence of critical thinking and participation, and the result is predictable: cynicism and indifference towards the public sphere.

This intriguing complexity is confirmed by a closer look at the ideology and practice of the middle classes. They can be either reactive, as is the case with a number of religious fundamentalist groups sharing neo-traditional values such as Santi Asok and Thammakai in Thailand and Al-Arqam in Malaysia, or proactive as a number of non-governmental organizations who according to McVey 'are almost always based on university-connected cadres. These NGOs have generally aimed at establishing or supporting moral communities whose boundaries do not mirror those of the state apparatus, which is seen as a source of repression rather than the font of legitimacy.' On the contrary, the appearance of NGOs might be interpreted as a response to the failure of the political establishment (i.e. the state) to dominate and control the vast changes which the states of Southeast Asia have launched; in particular the difficulty of adjusting to the rapid expansion and new priorities of their countries' middle classes, (McVey 1995: 7). Even though NGOs are generally speaking a middle-class phenomenon, they have shown a tendency to use their capacities to agitate for an empowerment of the underprivileged classes. Nevertheless, they cannot be regarded as social or political movements, 'even if they have the potential to precipitate them through the legitimation of class-based action' (Hewison and Rodan 1994: 258).

Attempts to define class in primarily cultural and social terms or to discount the idea of class divisions entirely divert attention away from the underlying social reality. Although the dominant Western discourse *per se* tries to label the current social upheavals in Southeast Asia as struggles for democracy, as if they were mostly concerned with political rights, reality attests that there are other dynamics at play which significantly challenge the version of (Anglo-Saxon) mainstream political science. Regardless of all else these labour- and resource-rich economies share two important problems: the unresolved agrarian question and conflicts in the rapidly changing social fabric. These structural aspects pose a challenge to the evolution of these societies and contribute to the importance of the development agency.

There is a well-known proposition by Gerschenkron that the later a nation enters industrialization, the more it requires guidance from above, meaning that the actual structure of the state is a determining part of the explanation as to why late developers have performed unevenly in the world economy (1962). This is confirmed by late developers such as Norway, Finland, New Zealand, Australia, South Korea, and Taiwan, which all share three basic features of development:

1. Successful agricultural modernization, resulting in not only increased productivity but also in a certain rural egalitarianism capable of creating an important home market for industrial goods
2. Strong linkages between the primary and secondary sectors, and more generally the creation of an industrial sector with, sooner or later, its own niche in the world market and competitive in certain limited areas;
3. A relatively effective and interventionist state apparatus playing a crucial role in the modernization of agriculture as well as in its effective articulation with industry. (Mouzelis 1995: 227–8)

This development-policy recipe has important implications in

Southeast Asia. The Philippines has one of the worst income-distributions in all of Asia, and the largest percentage of population in poverty in Southeast Asia. In spite of continued half-hearted agrarian reforms, the land tenure system continues to be one of the worst in the region. Unless there is land reform, which would take the poverty-ridden 70 per cent of the population residing in the countryside into a viable market, no amount of fiscal and monetary tinkering by structural adjustment programmes will produce sustained development. One of the problems with the unresolved land reforms in Southeast Asia is the continuing refusal by the World Bank and IMF to condition their loans on the basic redistribution of land, resources and wealth that would reverse growing inequalities (Broad et al. 1992).[6] Schmidt's and Somboon's contributions in chapters two and nine in this volume point to repeated failures in the region to implement comprehensive land reforms. In fact, this is the most pertinent problem and, as also emphasized in the end of this chapter, it creates a latent conflict potential involving the vast majority of the region's populations.

The importance of social tensions has so far been repressed ideologically by an almost hegemonic conservative liberal alliance between select authoritarian Asian leaders and their adherents in the West, which has vigorously promoted the idea of 'Asian values' to deflect popular discontent. According to Rodan, this has led to a situation where

> other divergent perspectives from within the non-government and grass-roots communities of course have less access to media to challenge the 'Asian values' line, and pursue agendas such as welfare, human rights and social justice issues as well as, in the case of some developmental NGOs, the establishment of greater participatory democracy and a shift in social power. (1995a: 14–15)

This exceptional situation where suppression of political and social rights exists in the name of culture is in fact similar to and resembles what Deyo has, in the Korean and Taiwanese contexts, termed 'development paternalism':

> The political and economic strategies of East Asian elites draw moral strength from two closely related sets of values. The first of these, centered on paternalism, invokes the moral authority of leadership that both defines and pursues national (vs. sectoral) interests through bureaucracy and public pronouncement. The second asserts the efficiency of such leadership through its proven material consequences for the public wealth. Economic development is the chosen measure of national welfare and thus a crucial basis for political legitimacy. Development paternalism, the composite of these two legitimating principles, justifies political exclusion and authoritarian rule as necessary for continuing high levels of growth. Alternative legitimating principles are effectively excluded from the moral domain of public discourse. (1989: 189)

The hegemony of 'developmental paternalism' in Southeast Asia has a strong influence on the nature of the emerging labour markets as well. While it is plausible to use the expression of Export-Oriented Industrializing regimes, they by definition demand either inclusionary policies towards labour, as has also been the case in South Korea, or exclusionary policies as evidenced in Taiwan; this ideological pattern is still basically different from the other NICs and Japan. As Robison emphasizes there is no evidence which shows the middle class, generally speaking, adjusting and conforming itself to the prevailing political regime, whether authoritarian or democratic. As pointed out by Schmidt and Drakakis, there are no immediate signs of a new social compact emerging anywhere, except for the social impact by labour and the underclass, which is also one of the major actors in South Korea after the softening of the authoritarian regime. These observations lead to another point of importance when the 'hard' Northeast Asian model is compared with its 'soft' equivalent in Southeast Asia.

The difference between Southeast and Northeast Asia is due not only to internal conditions, but is explained also by the external dimension. While Northeast Asia was submitted to the geopolitical attention of the United States, Southeast Asia can be said to have received a more pronounced geoeconomic interest from the developed world, especially after the neutralization of Asian socialism (see Hersh in chapter one). Thus, contrary to the experience of the East Asian NICs (except Singapore), the major impetus towards rapid industrialization in the Southeast Asian would-be NICs has primarily been an outcome of the search by core capital for low-wage labour, which has been accomplished through a transborder expansion of subcontracting networks originating in Japan (Arrighi et al. 1993), and later by capital exports and relocations of industries from the first-generation NICs. Paralleling this restructuring process, vast pools of low-wage labour in Vietnam and other East Asian latecomers (China) are increasingly being made available to international capital (Palat 1993: 18–19). This puts increasing competitive pressures on the would-be NICs, forcing them to replace labour and raise productivity. As usual the Philippines is the exception to this general picture; this is the only country where over the past two decades there has been rising labour productivity and at the same time a decline in real wages (Limqueco et al. 1989: 70).

Incomes in Malaysia and Thailand have expanded rapidly during the past two decades, although as emphasized by Fold and Wangel, the Malaysian government keeps wages down and rejects the implementation of a minimum-wage law in order to attract more foreign investment. The result is a highly segmented and polarized labour market with low increases in labour productivity. In the Thai case, employers have increasingly relied on short-term cost-cutting strategies to enhance product flexibility – involving quick adjustment capacities to changing market demands and flexibility in labour and other production costs. Deyo's study of Thailand reveals that the social consequences of these trends also increase labour casualization, sweatshop subcontracting, and widespread violations of minimum wage and social security laws. The decision to employ temporary labour has been followed by a number of cases of labour disputes or workers' efforts to establish labour unions (Deyo 1995: 132–3).

This is in a country where union density is very low, 'at roughly 6 percent overall, and it is even lower in the private sector' (Deyo 1995: 139). Furthermore, labour relation laws in Indonesia and Thailand exclude rural workers from labour legislation, thus rural workers do not have the right to form trade-union-type organizations. However, the relative absence of labour militancy in the region as a whole might be explained by the substantial labour surplus in the economies, except in Singapore and Malaysia in the 1990s, although repression by the state apparatus is another more subtle explanation.

Fear of losing jobs is not an exclusively Western phenomenon. In 1993 a wholly new trend emerged in Thailand, where more than 1,500 textile workers were laid off. Employers talk of up to 30,000 more job losses per year in a sector which currently employs some 800,000 Thai workers. Because of competition from Indonesia, Vietnam and China a number of companies have seen declining profits since 1989, and some have even experienced huge losses. According to an interview in *The Economist*, Yongkiat Tirachaimongkol, the director of finance from Thai Durable reported that, 'Thailand is no longer a good place to do labour-intensive work – The whole industry is struggling'(21 August 1993). The result is that newly-emerging trade unions, which have appeared since the fall of the Suchinda military regime, are protesting against lay-offs and the relocation of enterprises to other low-wage countries in the region.

This production-cycle or 'flying-geese' pattern of capital seeking new cheaper investment outlets for export manufacturing to foreign markets results in what Greenfield notes with regard to the new politics of exclusion of labour in Vietnam:

> Ironically the statist analyses of East Asian development undertaken by analytical Marxists and the progressive Left in the West now inform an agenda where the crushing of working class struggle and repression of the labour movement is implicit in this model. The trade union leadership itself has adopted the Singaporean model of trade unionism: business unionism and peaceful co-existence with the authoritarian-capitalist regime – a strategy by which they seek the political marginalization of the working class. (1994: 204)

According to Noerlund in chapter eight, most strikes in Vietnam occurred in foreign-owned enterprises, which confirms that the Singaporean strategy of containment and inclusion in state and party-linked unions is viewed as a more suitable way of dealing with labour conflicts.

However, since Vietnam has become a member of ASEAN, it is not surprising that this country has to adjust to the norms and labour codes of the other countries. All over the region this legislation has one primary function: to avoid conflicts in the labour market and attract increasing levels of foreign capital which again demands a docile labour force.

Understanding class and state in Northeast Asia: relevance for Southeast Asia

For more than a decade, the economic 'miracle'[7] of South Korea and Taiwan has been the object of a heated academic and political polemic. In the initial phase, the dispute was marked by the dichotomy between modernization approaches and dependency theory. Advocates of the former line of thought stressed culturalist determinancy, the dominant role of the market and the retrenched position of the state and its elites as major causes of these countries' rapid economic and social development (Pye 1985). Adherents of the latter position emphasized the constellation of favourable contingent conditions deriving from these countries' strategic geopolitical location (on the periphery of Asian Communism at the time), a world market relatively open for export-oriented models of industrialization, and the non-democratic nature of these states. In the dependency perspective, the significant aspects behind this type of growth process negate the liberalistic case of development and strengthen dependency assumptions (Evans 1987) that the 'Japanese model' is almost impossible to replicate in the hostile international environment of the 1990s (see also chapter one by Hersh).

In recent years, these two contradictory poles have more or less dissolved and given way to other interpretations. This tendency could be seen in a Japanese-financed World Bank study on the significance of economic growth policies in a number of East Asian economies (1993).[8] Even if the study could be seen as a platform for new policy guidelines it is essentially a disguised or simulated free-market approach (Wade 1990: chapter 1) or what Amsden calls 'market fundamentalism', to such an extent that it cannot prove its own major conclusion: 'It is quintessentially political and ideological' (1994: 627). It might as well be used indirectly as a new tool of control and domination to reinforce social clauses on trade, thus increasing state budgets in East and Southeast Asia on welfare and other non-accumulative expenditures.

In relation to the theme of changing class constellations in Southeast Asia, the NICs and Japan offer some unique features in their endeavours to combine rapid sustained growth with relative equal income distribution. Along with their pace of growth, the NICs and Japan have succeeded in combining repressive labour regimes and reducing income inequalities to such an extent that human welfare, and all the subordinate indices thereof, education, health, housing and so forth, have improved dramatically. On the other hand, these innovations have only been successful as a result of strong proactive state involvement in industrial restructuring, particularly in such areas as training, infrastructure, export promotion, research and development, and the creation of a strong supplier base for major industries. The historical background to this very unusual situation, which can only be said to be possible in societies which have experienced a fundamental revolutionary break with the past – for instance in China and Vietnam – is very well documented by a number of authors who carry further the arguments of Gerschenkron referred to earlier. In the case of South Korea: firstly, the position of the domestic landlords was undermined by land reforms after the end of Japanese colonialism; secondly, the domestic merchant capitalist class was weakened by state nationalization of the financial system; thirdly, no labour aristocracy emerged because of labour surplus and state repression. These factors left room for extensive state intervention and governance of the economy (See Cumings 1987; Evans 1987; Amsden 1989 and Wade 1990).

This has not been the case in Southeast Asian would-be NICs, where recent external pressures have resulted in neoliberal reforms of trade, industry and labour policy, which have discouraged enterprise reforms supportive of upgrading. The different hostile international climate of the 1990s has changed easy access to the US and EU markets, and in the Thai case, led to ' "market despotism" or, alternatively, autocratic internal flexibilization; imposed increased hardships on labour; and undercut the capacity of organized labour to contest new managerial strategies' (Deyo 1995: 142). It is also contrary to the strong repression of labour movements in Taiwan and South Korea which have been characterized by 'factory despotism' (Shin 1996: 13).

If there is a new course it may be stalled by recent efforts to dismantle the basic idea behind the whole debate: according to this line of thought, there is no such thing as *the* East Asian model of economic growth! This argument stresses that the four Tiger-countries (including Japan and China for that matter) display – and did already at the time of their economic take-off – vast differences in culture, political system, economic structure, constellation of social classes and position in the international division of labour (Shin 1996). This is not a new interpretation, however, as a number of recent studies have stressed the differences among the growth economies and raised substantial doubts about their level of deepening and sustainability (Arrighi et al. 1993; Gills 1993: 207). Nevertheless, the interesting task is to look for general policy or explanatory factors. This need not be the same as looking for a common 'model' or 'Asian system'.

However, the new 'simplicists' are discarding the idea that general policy lessons for promoting economic growth can be learned from the experiences in East Asia. Moreover, some even contest the significance and importance of sophisticated industrial strategies and selective protectionism, while arguing that these policies have not increased efficiency in the economies (Krugman 1994). According to this line of reasoning, economic growth has been a straightforward result of massive inputs of capital (due to a high savings ratio) and educated labour (due to increased educational level); in other words, a one-time change that cannot be continued once the economies are starved of capital and qualified workers.

One policy lesson cannot be derived from any sophisticated debate about state and market: 'If there is a secret to Asian growth, it is simply deferred gratification, the willingness to sacrifice current satisfaction for future gain' (Krugman 1994: 78). That may be so, but seemingly the result is that although Southeast Asian economies might continue to grow in the short-term, growing social tensions and a progressively reduced ability to compete with lower-cost producers will probably force a reconsideration away from neoliberalism to changes in economic strategies (Deyo 1995: 142). Furthermore, the region differs considerably from its Northeastern counterpart in terms of what *Newsweek* calls 'consumption-mad tigers' like Malaysia, Indonesia, and Thailand (Hirsh: 20 November 1995).

The question remains, however, as to how this incredibly rapid and comprehensive process of resource-mobilization was made socio-politically possible in the four East Asian countries. If it was not the result of sophisticated economic strategies, including selective industrial policies, the explanation needs to focus on the causal factors behind the socially and politically stable societies and the accompanying complex institutions and relations that pre-empted new class constellations from destroying or shaking the relative stability and making the massive mobilization of resources possible. Of course, US support for these military and authoritarian regimes played a significant role (see Hersh in chapter one).

Compared to the debate on the causes behind the economic growth of East Asian countries, studies of Southeast Asian countries have rarely taken the trouble to generalize the development process in model-terms. The obvious differences between the natural resource-rich countries in Southeast Asia – economic structure, ethnic composition, forms of regime etc. – seem to have forestalled such an approach.

What may appear to come closest to the debate on economic development in East Asia is the discussion of the 'flying geese' concept which was launched by Japanese scholars in the 1930s (Kojima 1977). Basically, this is the notion that first the East Asian and then the Southeast Asian countries follow Japan in the flight to economic and technological development by exploiting their comparative advantages in labour costs. Moreover, in so doing each country gradually adopts similar growth policies to those previously used by the countries in the upper economic echelon. Supporting the whole process is the flow of foreign (direct) investment in labour-intensive manufacturing industries in the late-comers; as countries move up the technological ladder and labour costs increase in the more advanced economies, there is more external capital available for cheap manufacturing in lower-cost countries.

Regardless of its apparent explicatory value, the 'flying geese' image is superficial and rejected by East and Southeast Asian scholars; although there may be some likeness in the efforts to emulate Japan (and South Korea), there are numerous differences related to policies as well as to internal and external conditions for industrialization and economic development in the Southeast Asian high-growth countries. Likewise, the notion of the 'flying geese' creates an image of a politically and economically united East and Southeast Asia that is far from being the case. Due to traditional and new rivalries in the region, extra-regional alliances, different business practices, and so forth, Southeast Asia offers a picture of great disparity. Furthermore the image is premised on unequal technological capacities and the ability of the state to wield 'direct and selective control over direct investment outflow to eliminate anti-trade investments' (Kojima 1977: 115).

Among the major differences between the first- and second-generation

Asian NICs is the increased penetration of global consumer culture and the difficulties of handling distributional issues by less authoritarian regimes within a changed international political economy. Thus, there is much more to explain than just simple similarities or differences between both generations of NICs' ability to mobilize resources in the 'Soviet fashion', i.e. to increase the I/C ratio and labour-force participation ratio and keep them at a high level.

It is worthwhile to note that the mobilization of resources in both first- and second-generation industrializing countries has until now taken place in periods of general growth, interrupted only by momentary stagnations in the process. This is presumably the reason why societal stability in the different countries has been maintained or easily restored after short-term unfavourable business cycles. It is difficult to say, though, what the economic, social and political consequences of a prolonged economic downturn due to a halt in the possibility of continued input-based growth would have been. According to one view it is highly questionable whether any of the Pacific-Asian states is able to replicate the so-called 'Japanese model' (see also Schmidt's contribution in chapter two). In fact, several indicators show that the East Asian NICs and Southeast Asian would-be NICs are experiencing a peripheralization which, paradoxically, may result in improving welfare for the mass of the inhabitants of the region, who have long been excluded from the benefits of economic growth, through a 'democratization of poverty'. The increasing social power of labour and support for human rights and democracy movements by core states and world financial institutions have eroded the legitimacy of authoritarian ruling elites to such an extent that a redistribution of income and political power in favour of the vast majority is a real possibility (Palat 1993: 20).

In order to explain the previous and prevailing societal structures as well as the economic and technological development of the Southeast Asian growth societies, more attention must be devoted to the issue of old, evolving and new classes and class constellations, not only in individual countries but also in the region.

Existing classes are being transformed and mobilized to enter into alliances with new classes, themselves the results of late-twentieth-century capitalist development. Class crystallization may take other forms and have other consequences than the traditional Eurocentric perception would suggest. Economic growth does not automatically foster a Western civil rights-type bourgeoisie prototype or a militant and outspoken working class. This notwithstanding, non-traditional countermeasures of different scopes and forms taken by oppressed classes and alliances are materializing on both the macro and micro levels in reaction to the political regimes.

As documented in this book, it is necessary to pay attention to these issues in order to understand the present societal dynamics in the countries/region and the political effects of the continuous production of inequality. Empirical sensitivity is needed to avoid sweeping regional generalizations that disguise important differences in the emergence and constellations of classes in these societies. This is of particular importance in the discussion of the roles of the middle classes and the working class.

Without doubt the importance of 'new' middle classes is due to the 'middle-classization' taking place in all the societies involved in the rapid economic transformation. The pressure from large segments of the population that can afford to take part in the consumption boom results in a situation where society's progress becomes proportional to the increased access to durable consumer goods and leisure services. As mentioned previously, this is a completely different situation from the one that prevailed for several decades in the first-generation NICs where frugality was the order of the day. Expectations about a better future in Southeast Asian countries will not keep down a strong aggregate demand for universal consumer goods; likewise the possibility of mobilizing resources (both capital and human) without immediate material rewards to established 'consumers' is non-existent.

The increase in income and opportunities for rising consumption have not been uniformly distributed among individuals and households within the middle classes. These – in essence – new classes have different positions in economic life, both in the production and reproduction spheres. Thus, reactions to different or dramatically changed living conditions depend on the actual material background of the different social groups that constitute the middle classes. The increasing attention paid to the middle classes related to the highly important question of political legitimacy should not crowd out important issues concerning the creation, oppression and incorporation of the working class(es) in the economic growth process, nor should the conditions of the rural 'proletariat' be ignored. As pointed out by Somboon and other authors in this volume, the means and scope of these efforts are significant to providing an understanding of the dynamics in these societies.

One of the outcomes of the development path in the Southeast Asian growth economies has been an increase in the numbers of low-paid jobs in the service sector. This can be seen in the domestic manufacturing sector to the extent that demand is covered by local production. Many of the new (service-related) jobs have no status whatsoever and the labour force in these positions apparently does not belong to either the traditional rural or the urban proletariat. They are, nevertheless, part and parcel of an evolution whereby new groups originating in the rural areas are incorporated into wage-relationships. As a result, the increasing fragmentation of the high wage-labour category, the working class, is an ongoing process that in some ways corresponds to the fragmentation of the middle classes. This evolution might be more rapid within the former case as technological development increasingly splits the production process into high-skill operations and activities which it is possible to carry out with poorly qualified labour. The lack of an adequate skilled work force within certain industries, particularly electronics, may actually hamper the further growth of manufacturing industry in the Southeast Asian generation of NICs as parts of production are (re)located to neighbouring countries with lower labour costs or more skilled labour. These havens include some of the region's 'high-tech islands'.

It is unquestionable that the economic success of the countries of Southeast Asia has contributed to the transition from dictatorship to 'soft authoritarianism' under the pressure of social forces together with the requirements of industrialization. The *quid pro quo* arrangements between the political regimes and the populations have depended on the ability of the economy to fulfil the ambitions of the political classes, the expectations of the middle classes, the living conditions of workers, and to a certain degree, the marginalization of the peasantry. The important question which arises at this juncture is whether this trade-off, which is the basis of relative stability, can be expected to continue unaltered, or whether it is liable to radical transformations. Although most of these states are authoritarian in character, albeit with a façade of elections and democracy, few have provided the mechanisms for peaceful successions of power for their highly personalized regimes. As yet, institutionalized rules for political changes are still weak and untried in most countries of the region.

In addition to putting internal pressures on the economic and political systems, the West has also started to pressurize Southeast Asia with its self-serving discourse on democracy and human rights. The motivation of the United States and Europe is not entirely unrelated to their own difficulties in the world economy. While Southeast Asian economic growth has been dependent on access to external markets and inflow of foreign investment, the Western world has become more wary of competition from this area. Many voices have been raised in Europe and the United States in favour of either some form of protectionism from low-wage producers or the more neoliberalist discourse of bringing Western production costs down so as to be competitive with these low-cost producers.

The dilemma of course for the second-generation NICs is that their comparative advantage is based on the correlation of low-cost production and socio-political stability. This is in fact a significant incentive for transnational corporations and foreign capital to invest anywhere in this age of globalization. As a reaction to Western criticism of their political systems the discourse of 'development paternalism' has played on the cultural differences between East and West, not only to legitimize the exclusion of labour from decision-making processes, but also in order to justify the lack of democracy in the countries of the region. In the view of Lee Kuan Yew or Mahathir Bin Mohamad, the debility of the West is due to its overemphasis on individualism instead of collectivism, on the welfare state instead of familism, and on moral laxity instead of discipline. While some of the arguments are self-contradictory they do serve to mobilize internal support for these regimes against external ideological pressure. Interestingly, even opposition groups have attacked the West for its interventionist stance in this area.

While the 'ideal-type' Western capitalism cannot be replicated – as originally implied by modernization theory – some questions are being raised by Asia-scholars as to whether we are here dealing with a '*sui generis*' form of industrial capitalism. The answer to this question will depend on the capability of these new high-growth economies to establish themselves as viable prototypes. The commonalities in these highly diversified societies are that, historically, civil society has been weak, the state has played a central role, and the bourgeoisies, local entrepreneurs, middle classes and 'the new rich' emerged from agrarian pre-capitalist and colonial bureaucracies and sometimes, from Communist Party rule (Robison and Goodman 1996: 4). The state in Southeast Asia may be compared more with that of the Japanese Meiji Restoration, albeit with the extremely important qualification that neither Japan nor Korea ever depended on foreign capital investment in their developmental paths.

Emerging social forces and changing relationships to the state

The best method for assessing the kind of capitalism the region is experiencing may be to look at different parameters, which, however, may overlap each other: a) economy, b) politics, c) security/military, d) culture, e) type of governance and f) stability and role of institutional succession, or the prospects of peaceful political change over time. While many of these aspects have been taken into consideration by the contributors to this volume they are not systematically treated, as some have put more emphasis on some factors while underplaying others.

Although the previous chapters have been principally directed at individual countries, an attempt to evaluate the prospects for the entire region can nevertheless be attempted. As mentioned above, Southeast Asia, being so differentiated, could not aspire to a role-model status. This ought to curb attempts at too encompassing generalizations. Although the degree of intra-Asian trade and investment has increased in later years with a corresponding reduction of dependence on the United States, there is yet little evidence of effective regional political cooperation. This is generally regarded as a negative sign as the area will in the coming decades have to face a number of challenges. Some of these are related to the economies of these countries. More sober evaluations of their prospects than official pronouncements have been voiced in recent years, suggesting that these societies might be close to the limit of their potential growth and be confronted with inflationary tendencies. On the political level, the semi-authoritarian systems of the countries of Southeast Asia might have to face intra-elite competition for power as well as demands of new groups, i.e. the rising middle classes, for inclusion in the decision-making process. The working classes will increasingly be mobilized to defend their positions, while the peasantry will put pressure on the internal distribution of the economic surplus. Transcending the contradictions between the traditional social categories it can be expected that youth and women, but especially ecological movements, will also make claims on the socio-economic pattern of development. An ultimate area of instability might be related to the emergence of military conflicts as sections of the political class look for 'external' enemies in order to mobilize support within their societies. This is strongly indicated by the latest trends, which show that East and Southeast Asia, including China, have become the largest market for world arms exports.

What distinguishes Southeast Asia from other developing regions around the world is the fact that the overwhelming majority of businesses have been in the hands of domestic Chinese minorities, and 'that indigenous entrepeneurs have lagged far behind them, both as private businessmen and as successful managers of state enterprises, despite massive government help in the form of protectionist and discriminatory regulations' (Mackie 1992: 162). Although this present book has focused on changing class constellations it has only touched briefly upon the role of big business and the petit bourgeoisie. New social forces have grown considerably outside the structures of the state. While the significance of supporting growth coalitions such as organized labour in both Indonesia and Thailand is recognized, new peasant movements and rural activism have come to the fore in Thailand, Malaysia and Indonesia as strong pressure groups resisting unevenness and intrusion of both global and local capital. Besides involvement in the question of community and land tenure rights and clashes surrounding eviction, these new organizations resist the rape of forests, blockading timber destined for export markets in Japan and the West, and oppose transnational mega-projects disguised as reforestation programmes like the eucalyptus plantation project in Thailand.[9] However, it remains to be seen whether these events are examples pointing to a substantial political transformation or whether they are simply the logical outcome of outside pressure from the Western economies and multilateral institutions to liberalize the political and economic contexts.

Jeffrey Winters' comments on poverty and equity in Indonesia could apply equally to the whole region, although with nuances and differences from country to country. Economic development in the region has brought staggering wealth for a few (especially the political elites, some groups within ethnic Chinese communities and, to a certain degree, the middle classes) amid widespread poverty. The privileged classes enjoy a glitzy cosmopolitan life and most of the basic freedoms that tend to accompany elite comforts. 'Meanwhile, the vast majority of the population lives in fearful silence, very mindful of the many occasions on which Suharto has demonstrated his willingness to deal swiftly and brutally with mass-based opposition.' Due to an artificially low officially stated poverty line the number of poor was 'only' 27 million in 1994; according to Winters, the figure should be closer to 75 million. The wealth gap is also much greater than is reflected in official estimates and World Bank publications, which give the impression that the Southeast Asian region as a whole, except the Philippines, is a role model for developmental success (1995: 420–4). These observations, which are admittedly hotly disputed, nevertheless point to the fact that national data on inequality and unevenness should be handled with care. As a matter of fact, during a field trip in 1993, one of the editors of this volume conducted interviews with several high-level policy-makers and planners in the treasuries and finance ministries of the region and got the same ironic message: 'We have one set of data for multilateral institutions and for official use, but the real data is kept in the safe'.

The conclusion which can be drawn from looking at the dynamism of Southeast Asia is one of awe in the face of the achievements – seen from an economic perspective. However, as an antidote to the general optimism concerning the future of the countries of the area, this volume should have made the case for a more sceptical approach regarding the resolution of the contradictions created by the favourable implementation of the high economic growth paths. There are too many unknowns to make too definite predictions. But it is certain that the countries of Southeast Asia as well as the region itself no longer remain the objects of history but are becoming movers in and shapers of world history. Concerning the question of whether a new social contract is emerging, the straightforward conclusion is that there are no signs of the emergence of a peaceful non-paternalistic method of resolving the problem of public goods in the region. The situation is fraught with social uncertainties, and other factors such as the form of the political regimes, restructuring of state apparatuses and, not least, political legitimacy play important roles in this regard. Other issues like personalized rule and succession in both Malaysia and, in particular, Indonesia point to changes in the near future in the correlation between political control and social organization. Although these countries have achieved a strengthened international position, the impact of actors external to these societies will still – in the view of the editors – influence the outcome of class segmentation and conflict patterns in the region.

1. 'In the hands of these neo-Marxists, the fairly unremarkable conceptual terrain of Marxism became a wonderland game of croquet played, as Koestler (1980: 63) once put it, with "mobile hoops." It meant among other things a deliberate turning away from economics and politics toward cultural and aesthetic critiques' (Kanth 1992: 199). On the brink of the new millennium, social sciences are increasingly being invaded by a wave of culturalization; theorizing is mostly concerned with compartmentalized and fragmented concepts and it has become very issue-focused. This is best illustrated by renewed scholarship about the so-called clashes of civilizations (Huntington 1993), and Asian versus Eurocentric monolithic values and cultures (Fukuyama 1992). For a good critique pointing to the ideological hegemony of attempts to impose a non-materialist terrain on to social sciences and policy debates in the East and Southeast Asian context, see Garry Rodan (1995a and 1995b).

2. It is not a proposal for a new theoretical paradigm (i.e. a set of axiomatic relationships that can be used to generate universal predictions of developmental outcomes) but rather some shared assumptions and hypotheses about the growing importance of social actors in development. See the discussion in Evans and Stephens (1988: 739–73).

3. For a critique of this interpretation in the Thai context see Schmidt (1996).

4. This has, in fact, been a salient feature of the strategy of the postcolonial state in Malaysia since 1969. The Malay statist capitalists have been ascendant since independence and, according to Jomo Kwame Sundaram, 'the more pronounced forms of class contention in contemporary Malaysia are outcomes of class contradictions generated by the rise of the statist bourgeoisie' (1988: 272).

5. See the contributions in McVey, and her own introduction (1993: 25).

6. See also Gustav Ranis (1989: 5 quoted from Broad, et al. 1992: 95 and 109 fn. 20).

7. The term 'miracle' is a misleading expression as it reduces the necessity for coherent and scientific explanations of very down-to-earth socio-economic and political processes.

8. Japan began openly to express disagreements with the World Bank in the late 1980s and

early 1990s on such issues as: whether financial policies should be subordinated to a wider industrial strategy; whether government should adopt measures aimed directly at promoting investment; whether these measures should be part of an explicit industrial strategy designed to promote leading industries of the future; whether directed and subsidized credit has a key role to play in promoting industry. For the background and a critique, see the contributions in Fishlow et al. (1994).

9. For a detailed review of the pressure of big business through political connections, transnational corporations and international aid agencies, as well as attempts by financial institutions to carry out plans for similar eucalyptus plantations working against the environment, ecology and land-rights of farmers in the ASEAN-4, with focus on Thailand, see Puntasen et al. (1992: 187–206).

Bibliography

Abercrombie, Nicolas and **John Urry** (1983) *Capital, Labour and the Middle Classes*, Allen & Unwin, London

Amsden, Alice (1994) Why Isn't the Whole World Experimenting with the East Asian Model to Develop?: Review of *The East Asian Miracle*, *World Development*, **22**(4) 627–33

Amsden, Alice (1989) *Asia's Next Giant – South Korea and Industrialization*, Oxford University Press, New York

Andreff, Wladimir (1993) 'The Double Transition from Underdevelopment and from Socialism in Vietnam', *Journal of Contemporary Asia*, **23**(4)

Arrighi, Giovanni, Satoshi Ikeda and **Alex Irwan** (1993) 'The Rise of East Asia: One Miracle or Many?' in Ravi Arvind Palat (ed.) *Pacific-Asia and the Future of the World System*, Greenwood Press, Westport, Conn.

Beresford, Melanie (1993) 'The political economy of dismantling the "bureaucratic centralism and subsidy system" in Vietnam', in Kewin Hewison, Richard Robison, and Garry Rodan (eds) *Southeast Asia in the 1990s. Authoritarianism, Democracy and Capitalism*, Allen & Unwin, Sydney

Booth, David (1994) 'Rethinking Social Development: An Overview', in David Booth (ed.) *Rethinking Social Development. Theory, Research and Practice*, Longman, Harlow

Booth, David (1985) 'Marxism and Development Sociology: Interpreting the Impasse', *World Development*, **13**(7)

Broad, Robin, John Cavanagh and **Walden Bello** (1992) 'Sustainable Development in the 1990s', in Chester Hartman and Pedro Vilanova (eds), *Paradigms Lost. The Post Cold War Era*, (the Transnational Institute, Amsterdam and the Institute for Policy Studies, Washington DC), Pluto Press, London

Corbridge, Stuart (1994) 'Post-Marxism and post-colonialism: the needs and rights of distant strangers', in David Booth (ed.) *Rethinking Social Development. Theory, Research and Practice*, Longman, Harlow

Cumings, Bruce (1987) 'The Origins and Development of the Northeast Asian Political Economy: Industrial Sectors, Product Cycles, and Political Consequences', in Frederic C. Deyo (ed.) *The Political Economy of the New Asian Industrialism*, Cornell University Press, Ithaca, NY

Dahrendorf, Ralf (1959) *Class and Class Conflict in Industrial Society*, Stanford University Press, Stanford, California

Deyo, Frederic C. (1995) 'Capital, Labor, and State in Thai Industrial Restructuring: The Impact of Global Economic Transformations', in David A. Smith and József Böröcz (eds) *A New World Order? Global Transformations in the Late Twentieth Century*, Greenwood Press, Westport, Connecticut

Deyo, Frederic C. (1989) *Beneath the Miracle – Labor Subordination in the New Asian Industrialism*, University of California Press, Berkeley

Deyo, Frederic C. (1987) 'Coalitions, institutions, and linkage sequencing – toward a strategic capacity model of East Asian development', in Frederic C. Deyo (ed.) *The Political Economy of the New Asian Industrialism*, Cornell University Press, Ithaca, NY

Doner, Richard F. (1991) 'Approaches to the Politics of Economic Growth in Southeast Asia', *Journal of Asian Studies*, **50**(4) November

Doner, Richard F. (1992) 'Limits of State Strength. Toward an Institutionalist View of Economic Development', *World Politics*, **44**(3) April

Economist, (21 August 1993) 'How Cheap can You Get?'

Evans, Peter and **John D. Stephens** (1988) 'Development and the World Economy' in Neil J. Smelser (ed.) *Handbook of Sociology*, Sage, Newbury Park, California

Evans, Peter (1987) 'Class, State, and Dependence in East Asia: Lessons for Latin Americanists', in Frederic C. Deyo (ed.) *The Political Economy of the New Asian Industrialism*, Cornell University Press, Ithaca, NY

Fishlow, Albert, Catherine Gwin, Stephan Haggard, Dani Rodrik and **Robert Wade** (1994) *Miracle or Design? Lessons From the East Asian Experience*, (Foreign Policy Essay No. 11), Overseas Development Council, Washington DC

Fukuyama, Francis (1992) 'Capitalism and Democracy: The Missing Link', *Journal of Democracy*, **3**(3), July

Gerschenkron, Alexander (1962) *Economic Backwardness in Historical Perspective. A Book of Essays*, Harvard University Press, Cambridge, Mass.

Gills, Barry (1993) 'The Hegemonic Transition in East Asia: A Historical Perspective', in Stephen Gill (ed.) *Gramsci, Historical Materialism and International Relations* (Cambridge Studies in International Relations: 26), Cambridge University Press, Cambridge

Greenfield, Gerard (1994) *The Development of Capitalism in Vietnam*, Socialist Register, Merlin Press, London

Hewison, Kewin and **Garry Rodan** (1994) *The Decline of the Left in Southeast Asia*, Socialist Register, Merlin Press, London

Hewison, Kewin, Garry Rodan and **Richard Robison** (1993) 'Introduction: Changing Forms of State Power in Southeast Asia', in Kewin Hewison, Richard Robison, and Garry Rodan (eds), *Southeast Asia in the 1990s. Authoritarianism, Democracy and Capitalism*, Allen & Unwin, Sydney

Hirsh, Michael (1995) 'Which Asian Model?' *Newsweek*, 20 November

Huntington, Samuel P. (1993) 'The Clash of Civilization?', *Foreign Affairs*, **72**(3)

Kanth, Rajani (1992) *Capitalism and Social Theory. The Science of Black Holes*, M. E. Sharpe, Armonk

Koestler, Arthur (1980) *Bricks to Babble*, Picador, London

Kojima, Kiyoshi (1977) *Japan and New World Economic Order*, Croom Helm, London

Krugman, Paul (1994) 'The Myth of Asia's Miracle', *Foreign Affairs*, **73**(6) November/December

Laothamatas, Anek (1992) 'The Politics of Structural Adjustment in Thailand: A Political Explanation of Economic Success', in Andrew MacIntyre and Kanishka Jayasuriya (eds) *The Dynamics of Economic Policy Reform in South-east Asia and the South-west Pacific*, Oxford University Press, Singapore

Limqueco, Peter, Bruce McFarlane and **Jan Odhnoff** (1989) *Labour and Industry in ASEAN*, Journal of Contemporary Asia Publishers, Manila and Wollongang

MacIntyre, Andrew and **Kanishka Jayasuriya** (1992) 'The Politics and Economics of Economic Policy-Reform in South-east Asia and the South-west Pacific', in Andrew MacIntyre and Kanishka Jayasuriya (eds) *The Dynamics of Economic Policy Reform in South-east Asia and the South-west Pacific*, Oxford University Press, Singapore

Mackie, Jamie (1992) 'Changing Patterns of Chinese Big Business in Southeast Asia', in Ruth McVey (ed.) *Southeast Asian Capitalists*, Cornell Southeast Asia Program, Ithaca, New York

McVey, Ruth (1995) 'Change and Continuity in Southeast Asian Studies', *Journal of Southeast Asian Studies*, **26**(1) March

McVey, Ruth (1992) 'The Materialization of the Southeast Asian Entrepeneur', in Ruth McVey (ed.) *Southeast Asian Capitalists*, Cornell Southeast Asia Program, Ithaca, New York

Mouzelis, Nicos (1995) 'Modernity, Late Development and Civil Society', in John A. Hall (ed.) *Civil Society. Theory, History, Comparison*, Polity Press, Cambridge

Paix, Catherine (1993) 'The Domestic Bourgeoisie: How Entrepeneurial? How International?', in Garry Rodan (ed.) *Singapore Changes Guard. Social, Political, and Economic Directions in the 1990s*, Longman Cheshire, Melbourne

Palat, Ravi Arvind (1993) 'Introduction: The Making and Unmaking of Pacific-Asia', in Ravi Arvind Palat (ed.) *Pacific-Asia and the Future of the World System*, Greenwood Press, Westport, Conn.

Puntasen, Apichai, Somboon Siriprachai and **Chaiyuth Punyasavatsut** (1992) 'Political Economy of Eucalyptus: Business, Bureaucracy and the Thai Government', *Journal of Contemporary Asia*, **22**(2)

Pye, Lucien W. (1985) *Asian Power and Politics. The Cultural Dimensions of Authority*, Belknap Press of Harvard University Press, Cambridge, Mass.

Ranis, Gustav (1989) *The Philippines, the Brady Plan, and the PAP: Prognosis and Alternative*, update of January 1989 report to the US Agency for International Development, unpublished paper, Yale University, May

Robison, Richard and **David S. G. Goodman** (1996) 'The New Rich in Asia: Economic Development, Social Status and Political Consciousness', in Richard Robison and David S. G. Goodman (eds), *The New Rich in Asia. Mobile Phones, McDonalds and Middle-Class Revolution*, Routledge, London and New York

Rodan, Garry (1995a) *Theoretical Issues and Oppositional Politics in East and Southeast Asia*, Working Paper No. 60, Asia Research Centre, Murdoch University, Perth, Western Australia

Rodan, Garry (1995b) *Ideological Convergences Across East and West: The New Conservative Offensive*, Working Paper No. 41, Research Centre on Development and International Relations, Aalborg, Denmark

Schmidt, Johannes Dragsbaek (1996) 'Paternalism and Planning in Thailand: Facilitating Growth without Social Benefits', in Michael J. G. Parnwell (ed.) *Uneven Development in Thailand*, Avebury, Aldershot

Shin, Kwang-Yeong (1996) *Characteristics of the East Asian Economic System: Authoritarian Capitalism and the Developmental State*, Working Paper No. 52, Research Centre on Development and International Relations, Aalborg, Denmark

Smelser, Neil J. (1994) 'Sociological Theories', *International Social Science Journal*, **139**, February

Subangun, Emmanuel (1989) 'Public Policy and Political Economy for the Low-income Groups in Indonesia', in Helmut Kurth (ed.) *Economic Growth and Income Distribution*, (Friedrich Ebert Stiftung), Moed Press, Quezon City, Philippines

Sundaram, Jomo Kwame (1988) A *Question of Class. Capital, the State, and Uneven Development in Malaya*, Monthly Review Press, New York

Thrift, Nigel and **Peter Williams** (1987) *Class and Space. The Making of Urban Society*, Routledge & Kegan Paul, London and New York

Wade, Robert (1990) *Governing the Market. Economic Theory and the Role of Government in East Asian Industrialization*, Princeton University Press, Princeton, New Jersey

Winters, Jeffrey (1995) 'Suharto's Indonesia: Prosperity and Freedom for the Few', *Current History*, **94**(596) December

World Bank (1993) *The East Asian Miracle. Economic Growth and Public Policy*, Oxford University Press, New York

Wright, Erik Olin (1985) *Classes*, Verso, London

Index